# Cross-Country Flying

TAB
PRACTICAL
FLYING SERIES

# Cross-Country Flying

*Jerry A. Eichenberger*

## TAB Books
### Division of McGraw-Hill

New York  San Francisco  Washington, D.C.  Auckland  Bogotá
Caracas  Lisbon  London  Madrid  Mexico City  Milan
Montreal  New Delhi  San Juan  Singapore
Sydney  Tokyo  Toronto

The materials presented in this book are subject to constant change and revision. The reader should not use any of the aeronautical charts included herein for navigational purposes, or depend upon quoted sections of the Federal Aviation Regulations or *Airman's Information Manual*. This book is presented for instructional purposes only, and no person or entity connected with it assumes any liability for the correctness of any of the material herein, or for the failure to include any material omitted from this book.

© 1995 by **TAB Books.**
TAB Books is a division of McGraw-Hill, Inc.

pbk      5 6 7 8 9 10 11 FGR/FGR  9 9 8
hc    1 2 3 4 5 6 7 8  9 10 FGR/FGR  9 9 8 7 6 5 4

Product or brand names used in this book may be trade names or trademarks. Where we believe that there may be proprietary claims to such trade names or trademarks, the name has been used with an initial capital or it has been capitalized in the style used by the name claimant. Regardless of the capitalization used, all such names have been used in an editorial manner without any intent to convey endorsement of or other affiliation with the name claimant. Neither the author nor the publisher intends to express any judgment as to the validity or legal status of any such proprietary claims.

**Library of Congress Cataloging-in-Publication Data**

Eichenberger, Jerry A.
   Cross-country flying / by Jerry A. Eichenberger. —4th ed.
      p.  cm.
   Rev. ed. of: Cross-country flying / R. Randall Padfield. 3rd. ed.
   1991.
   Includes index.
   ISBN 0-07-015076-1    ISBN 0-07-015077-X (pbk.)
    1. Cross-country flying.   I. Padfield, R. Randall.  Cross-country
  flying.  II. Title.
TL711.L7E35    1994
629.132'5—dc20                        94-19578
                                            CIP

Acquisitions editor: Jeff Worsinger
Editorial team: Robert E. Ostrander, Executive Editor
              Norval Kennedy, Editor
              Elizabeth J. Akers, Indexer
Production team: Katherine G. Brown, Director
              Lisa M. Mellott, Coding
              Wanda S. Ditch, Desktop Operator            PFS
Designer: Jaclyn J. Boone                                015077X

To the memory of Richard A. Moss

Dick Moss was the living definition of an aviator. I had the pleasure of meeting him when I was a young private pilot with the ink still wet on my temporary license. Under Dick's guidance and tutelage, I became a commercial and multiengine airplane pilot and a certificated flight instructor.

He was a fully rated pilot and instructor who also held an airframe and powerplant mechanic certificate with inspection authorization. When a pilot was schooled under Dick, the student learned how to fly like a pupil of Mozart would learn music. Dick Moss's kind of instruction is sorrowfully rare.

Dick produced a complete pilot who knew not only how to get an airplane from one place to another, but who was also imbued with a knowledge of the machine and how it worked. I thank him for almost all of what I know about the art that has been in one way or another the focus of almost all of my adult life.

After I had a few thousand hours in my logbook, I was still amazed at how an airplane became an extension of not only his hands and feet, but an extension of his senses as well. I remember the day when he climbed into a T-6 for the first time and proceeded to fly it as though he had years of experience with it. He ended that flight with a picture-perfect three-point landing onto a 3,600-foot narrow runway after masterfully slipping it over the power lines at the end of the runway. If anyone of the rest of us could even approach his insight and skill, we'd be among the best who ever took a control stick in our hands.

Rest in peace dear friend, and enjoy the new set of wings that you have been given.

# Contents

# Introduction

PILOTS FLY GENERAL AVIATION AIRPLANES FOR MANY REASONS. SOME OF us seek only recreation in the form of sport flying. Others see an airplane purely as a mode of transportation for either business or personal travel. A few aviators are into specialized flying, such as aerobatics, agricultural application, and the preservation and flying of warbirds.

A recent survey of general aviation pilots that was conducted by a major aviation publication disclosed that the reason the majority of all pilots were attracted to flying is the challenge of mastering a complicated machine and its operation. Why we fly isn't as important as doing it properly and continuing to learn and perfect the skills that keep aviation and its pilots as safe as possible.

Airplanes come in almost as many variations as the people who fly them. General aviation is loosely defined as all forms of aviation except the military and airlines. By that definition, our industry encompasses aircraft from the smallest and least complex homebuilt airplane up to and through the largest and most complex business jets: from Kitfox to Gulfstream.

Another component of general aviation is the helicopter industry. While most helicopters have been developed as a result of military needs, that tendency is on the wane, and many rotary-wing aircraft are now on the market that were designed from inception as civilian machines.

But there is one commonality among virtually all of the aircraft and uses that make up general aviation. Whether a pilot is flying a warbird to the next airshow, an applicator is repositioning the crop duster to the next job a few counties away, a business person is on the way to a meeting, or the sport aviator is just flying around on a weekend afternoon looking for a place to eat lunch, all are doing the same thing: cross-country flying.

Cross-country travel can take on different attributes according to the conditions facing the pilot. Some trips are flown under instrument flight rules, but the great majority occur under visual flight rules. The same route might be being flown simultaneously by a Citation up in the flight levels and a Cessna 172 only a few thousand flight above the ground.

It's possible to fly across our continent without ever turning a radio on. Other pilots might not be able to see anything outside the cockpit to know exactly where they are at a given time, depending instead on the magic of modern avionics for navigation to guide them through the rain and the clouds to a destination hundreds of miles away.

*Cross-Country Flying* cannot and will not try to address all of these possibilities. This book will concentrate on VFR cross-country flight in typical piston single-engine airplanes at the altitudes normally flown in them and will delve into the conditions that often confront the pilot. Neither will it cover helicopter cross-country operations as a separate subject because cross-country helicopter flying is very similar to doing the same thing in an airplane, except for the unique operational characteristics of rotary-wing aircraft.

Stepping up from VFR flight into IFR operations is an entire subject in itself; therefore, IFR flying will not be the subject of much attention in *Cross-Country Flying* because justice to it could not be done in a single volume that emphasizes VFR flying.

I certainly hope you enjoy these pages and learn a little from the stories and ideas conveyed. If you come to love flying or already have, you'll know the truth of the caption that accompanies one of the more popular pieces of aviation art: "Once you know how to fly, you will understand every pilot's sense of wonder."

# 1
# The many faces of cross-country flight

CROSS-COUNTRY FLIGHT HAS NO PRECISE DEFINITION. WITHIN THE GENERAL aviation world, it is usually used to mean a flight from one place to another, which requires some form of navigation. Cross-country flying generally means aviating some distance from the starting point to the destination, which is far enough that the pilot can't simply ignore navigation; the two airports are far enough apart that he is out of sight of one of the two of them during enough of the trip that he has to use some form of navigation to get from the place of beginning to the eventual destination.

In the Federal Aviation Regulations (FARs), the *Airman's Information Manual* (AIM), and other government publications, there are numerous references to cross-country flight. Most of these appear in Part 61 of the FARs, which deals with the requirements that a candidate must satisfy in order to obtain the various pilot certificates and ratings.

In the sections dealing with the aeronautical experience necessary to qualify for a private pilot certificate, the applicant must present evidence to the examiner that she has logged at least 10 hours of cross-country flight. Each flight must involve a landing at a point at least 50 nautical miles from the original departure point; one of the flights must be at least 300 nautical miles in total length with landings at a minimum of three points.

Additionally, during this "long cross-country" the student must land at one or more airports not less than 100 nautical miles away from the original departure point. So the Federal Aviation Administration's (FAA) regulatory scheme for licensing pilots deems cross-country operations to be flights of a minimum of 50 nautical miles in length.

When I was going through this process in the 1960s, the FARs considered 25 miles as the minimum distance for a trip to qualify as a cross-country. Those earlier regs were written in the days of the J-3 Cub and Aeronca Champ, when trainers cruised at speeds around 70 knots. Going that slow, it took a while to fly even 25 miles. Modern trainers have cruising speeds at least 30 percent faster and maybe 40 percent faster; therefore, the regulation writers increased the required distance to 50 nautical miles some time ago.

Airplanes come in all sizes and are designed for a multitude of purposes, but for the great majority of pilots and airplane designers, aviation is a form of transportation. Specialty aircraft include trainers, crop dusters, and aerobatic airshow and competition airplanes, but the primary purpose of almost all airplanes is to move people or cargo from one place to another at the least cost consistent with the performance parameters of the airplane involved and with an acceptable degree of safety.

One inexorable fact about flying is that the cost of operation of any given airplane is dependent upon its performance and load-carrying capabilities. A Boeing 747 costs

**Fig. 1-1.** *The beginning of a typical cross-country flight. The airplane shown is a Cessna Cardinal RG.*

thousands of dollars per hour to fly, while a two-place airplane such as a Cessna 150 can be operated for fewer than $50 per hour. The 747 carries hundreds of people across oceans at globe-shrinking speeds; the 150 carries two people at just over 100 miles per hour. Both engage in the same mission, which is carrying folks from point A to point B: cross-country flying.

In the realm of general aviation, cross-country flights have two primary purposes. First, a good number of flights are purely recreational, either for day trips, vacations, or training. The second type of cross-country flight is for a business purpose, transporting the pilot or the passengers or all occupants to another place to conduct business and return home. There is a significant difference between the two.

Recreational cross-country flying, especially of the day-trip variety, seldom involves any firm schedule. The pace is leisurely; if the trip has to be postponed a day or two for weather or other considerations, it is usually of little importance. A business flight is another matter. There is the sales presentation to make, the scheduled meeting to attend, or any number of reasons why the pilot feels that the trip must arrive on time.

Business pilots suffer from pressures that don't normally intrude into the flying routine of the pilot who uses an airplane only for pleasure trips; I've been there and know the difference. Meetings with consultants or a board of directors, depositions, court appearances, and similar events are scheduled well in advance; the other participants aren't particularly sympathetic to cancellations or late arrivals by a general aviation pilot. They usually fume, "The airlines got here, why didn't you?"

The business pilot has a life of flying in pursuit of his living. If he doesn't make the trip or arrives late, irretrievable financial loss is often the result. But the problem is that pressing on into weather that neither he nor his airplane can handle might well result in a totally irretrievable loss: a fatal accident. These aviators are business people first and pilots second. Flying the airplane is secondary because it merely provides the means to get somewhere to earn a living.

A couple of old adages say quite a bit to the VFR business pilot who has a tight schedule. "If you've got time to spare, go by air." The other goes like this, "If you must arrive, take the airlines or drive." All general aviation pilots would be wise to heed this advice in the face of planning a cross-country flight that their better judgment tells them ought to be postponed or canceled.

A pilot flying on a business mission is often thinking about the business purpose behind the flight, planning for what will go on when she arrives, not totally devoting every thought to the conduct of the flight. This division of attention has led more than one business flight into disaster. It's a plague that requires constant vigilance on the part of all pilots who fly on business trips. If the pilot isn't instrument-rated and flying an appropriately equipped airplane, the utility of the airplane as a business traveling tool will be compromised. The wise know these limits of themselves and their machines; the foolish press on at any cost, and the cost is often to themselves as well as their pocketbooks.

A good many cross-country flights still take place in older airplanes that have minimal, if any, onboard electronic aids to navigation. A trip without "black boxes" can be one of the most enjoyable that any pilot can make, if there is enough time to deal with all of the uncertainties inherent in this kind of flying.

The rash of hand-held *global positioning system* (GPS) receivers now on the market has made it possible to have the most precise of modern navigation equipment available in the most primitively equipped aircraft. Hand-held communications radios, when hooked up to an external antenna that can be mounted on virtually any aircraft, enable the old classic Champs and Cubs to use controlled airports right along with trainers of more recent vintage.

Regardless of the reason for the cross-country flight and the wide variation in pilot skills and aircraft capabilities, there is an infinite scope of flying involved. Some flights will be over the flat terrain of the midwestern plains; others will take the pilot into the perpetual reduced visibility of the industrial Northeast. A route might force the pilot to either accept the risks inherent in crossing larger bodies of water, such as one of the Great Lakes, or endure the extra time required to fly around the periphery. A flight might span the expansive swamps of southern Georgia and northern Florida, or crest 3-mile-high mountains in the West.

Vast deserts are crossed in the Southwest, and pilots see seemingly endless forests and isolated lakes of the northern states. Summer thunderstorms and winter icing present potentially lethal hazards. The winter season features reduced hours of daylight that require many cross-country flights to either begin early or end late in the dark of night. If a pilot intends to fly over any or all of these earthly variables, there is no end to the educational process that should be a lifelong undertaking.

No two cross-country flights are ever alike. Their features come in an endless variety of terrain, winds, weather, and schedules, never to be exactly duplicated again. The key to the successful completion of all flights lies in knowledge and preparation.

Most things worth doing in life require some degree of planning, and cross-country flying is no exception. The salesperson is best equipped to face a prospect if armed with as much information about the person as can be garnered. The trial lawyer had better know her client's case as well as the evidence that will be presented by the opposition.

The pilot who is setting off on a cross-country flight also needs to be forearmed with the best information available, such as the route to be flown, the weather, the performance capabilities of the airplane to be flown on the trip, and whatever else is to be encountered. The winds will influence the altitude at which the flight will be made, the time en route, and the need to make or omit a refueling stop or two.

Humidity and temperature can rule out the use of certain airports that might be just fine if the day were cooler or drier. Turbulence can turn a pleasant flight into bone-jarring misery, and an encounter with low clouds and rain will make the VFR pilot land and wait for conditions to improve. A forecast of lousy weather might force the pilot to cancel the flight and opt for other transportation.

Almost all pilots like challenges because the adventure of aviation is often the factor that motivates a person to learn to fly in the first place. A pilot with the right attitude approaches cross-country flying as the quantum of know-how and judgment, and using them is an enjoyable challenge. The sensible aviator assesses each flight beforehand and accepts only those flights that can be safely concluded.

Because this book will concern itself with VFR cross-country flying and not attempt to delve into the world of IFR operations, we'll take a look at a number of different flights and the conditions that the pilot faced. Most of the trips will involve a pilot or pilots well known to your author, so the stories are accurately related. First is an examination of some of the preliminary knowledge, choices, and tasks involved in a pilot's gaining an understanding of this subject of cross-country flying.

# 2
# Airports

UNLESS YOU FLY HELICOPTERS OR HOT-AIR BALLOONS, YOUR FLIGHTS WILL originate from and, if everything goes as planned, terminate at an airport. Like the human face, airports come in all varieties. Some serve airliners with 2-mile-long runways and are a beehive of activity. Others are sod strips that are nearly impossible for the first-time visitor to find and might not see a transient aircraft in days or even weeks. Some airports have control towers, but most do not.

In order to land at airports in some localities, the pilot has to deal with high-density traffic areas, positive controlled airspace, and complex regulations concerning operations in that airspace. Other destinations won't involve these considerations but might present other challenges. A short runway, high elevation, power lines or other obstacles, and rising terrain create special problems for a pilot to recognize and overcome.

Because we've already assumed that a cross-country flight will involve operations at a minimum of two airports, let's recognize another obvious fact. You must take off from the airplane's present location, but the choice of where you will land is quite another facet of cross-country flying. In order to make that choice, it's helpful to know some of the conventional wisdom about airports and operations at the various types.

The first determinate in the choice of a destination airport is obvious; how close is the field to your ultimate stop? If a cross-country flight is being made in order to transport you somewhere that you need to go, it doesn't make much sense to choose a destination airport that isn't convenient to the place you need to be. Assuming that you're headed to a major city, chances are that you will probably have a choice of which airport to plan to use upon your arrival into that area.

If your plans take you to a smaller town or village, the option might be decided for you if only one airport is close to your eventual stop. Even this situation demands that a choice be made; if the airport doesn't fit the airplane's performance requirements, or if your flying expertise doesn't mesh with airport operations, you'll have to plan to use a different field. There is no such thing as a flight that has to be made or a landing that has to be accomplished, absent an emergency landing.

## SOME PRELIMINARY CONSIDERATIONS

What kinds of airports do you like to use? Where do you feel comfortable with your airplane's needs and your proficiency? What traps might lurk at any given airport ready to catch the unwary or ill-prepared pilot? Lastly, what airport is convenient? Is there a choice? Let's take a look at these questions and expand upon the considerations that are necessary to answer them.

### The comfort factor

Not all pilots feel "comfortable" about using *all of the airports* in the United States at which their airplanes could be safely operated. The percentage of pilots that is comfortable flying at any airport probably represents a noticeable minority. Most pilots trained in the last 20 years or so have not flown at grass or other unprepared fields. Pavement is the norm, and a hard-surface runway is available now at all but the remotest locations.

Pilots who shy away from a sod field just because it is sod often deny themselves a convenient place to end a flight. These pilots also miss out on some of the last remnants of what general aviation once was. But sod strips aren't all the same. You cannot make the assumption that any grass field that is long enough for your airplane will be otherwise suitable, which is usually a safe assumption to make about the surface condition of almost all paved runways.

First of all, if you have never operated your airplane from a sod strip, don't wait until a cross-country flight into an unknown airport to try it. If you have a good idea of the minimum runway length needed for your airplane on a paved field, don't think for a minute that it will be necessarily sufficient on grass.

One of the biggest determiners of takeoff distance on sod is the height of the grass. If it's mowed down close and not too thick, the takeoff distance won't be much longer than what you'd experience at a hard-surface airport. If the grass is long and thick, you might be surprised at the increased length of the takeoff run before your airplane reaches flying speed. If the runway length looks even slightly marginal, you'd better stay away from the grass field unless you know in advance what the surface conditions are.

In a similar light, some grass strips are built to drain well, while others are pure mud holes in wet weather. Be careful if it's springtime anywhere or wintertime in the moderate climes when wet weather might be likely. Getting stuck in the mud is not only embarrassing, but nosewheels and propellers don't like the sudden lurching stop that a soft or muddy field can provide. Your landing roll won't be long; it might be a landing slide. If in doubt, get the phone number of the airport from an airport directory and call ahead.

Be aware that most modern airplanes aren't really designed to be operated off grass strips with any frequency. Your airplane probably has a nose tire that is smaller than the main gear tires. A skinny tire on the nose makes a neat little knife that slices down into wet or soft surfaces, making it all the easier to get stuck, or to break or bend the landing gear. A small nose tire might not provide enough propeller clearance from the ground if the runway's surface isn't very smooth. Be prepared. Know the field's condition, and have fun at a field that is in good condition.

Find out how smooth and level the sod is. I have landed on washboards that some airport operators had the nerve to call runways. If you've got a tricycle-gear airplane with either a small nose tire or minimal ground clearance for the propeller, you shouldn't chance a rough-field landing. Pilots are blessed with plenty of grass runways throughout the United States that are as smooth as a baby's skin and as level as a pool table. To my thinking, no other landing is as much fun or as pleasant as one made on a nicely prepared and maintained grass strip.

Tire wear will be lessened, and you won't need nearly the same amount of braking, if any, as on a paved runway. Remember that unless you know the airport or find out about it in advance, it is impossible to know what you will encounter upon landing, and then it might be too late.

Some tricycle-gear airplanes do a much better job of tolerating less than ideal runways than do others. Most of the lighter Cessnas like the 152 and 172 can take a more abusive runway in stride. A Mooney might pose a real problem with its low-slung undercarriage and minimal propeller clearance. If you fly one airplane regularly, you'll soon get a good idea of its abilities and needs. If you fly many different airplanes, be a little extra careful before operating at unpaved airstrips.

What other kinds of airports fit into your comfort zone? Assuming that sod isn't for you or you can't obtain information about a potential landing site, you're limited to a hard-surface airport. But there are as many varieties of those as there are places to go. Some airports with a hard-surface runway are nothing more than unattended fields with no facilities or support services or even a telephone. Large airports, such as O'Hare or JFK, are there if you need them and if you are willing to jump through the hoops of the reservation system and costs incurred to land at them.

It's probably a fair estimate to say that a general aviation cross-country flight terminates at a paved airport with a *fixed-base operator* (FBO) doing business on the field, ready to sell you fuel and supplies. Services that include aircraft maintenance are probably available from the FBO. This kind of airport is a good choice, unless winds are unusually strong and there is only one runway. Proficiency in crosswind landing techniques and practice to maintain that proficiency would probably allay any concerns about operating at a field with one runway.

The United States has about 5,500 public-use airports, and about 4,800 of them have runways at least 6,000 feet long. Counting all of the biggest airline airports, there are just 687 airport control towers in the country (1992 data). It's easy to see that there is a plethora of good, paved airports across the nation from which to choose. If the medium-size hard-surface airport (most of which have multiple runways) fits into your preferred operational mode and the needs of your airplane, there are plenty of them to serve you in all but the remotest of places.

In a short while, we'll discuss flying from and into airports that lie within some kind of airspace that demands communication with air traffic control (ATC). Remember that in all but the busiest terminal areas, the city airport that serves airline traffic is usually a good general aviation destination because you'll probably find the services that you might need, such as fuel, a rental car, weather reporting and briefing services, and a place to either tie down or hangar the airplane during the visit.

Cross-country flying demands proficiency in more of the total spectrum of flying skills than local flying. If your trip is planned for more than 100 miles or so, you might well fly through or into a different air mass than the air over your departure point. We'll spend a good bit of time in chapter 4 talking about the considerations that come into play when planning a cross-country flight, particularly those involving weather. If you see on your chart that there is a nice airport at your destination with a single runway that is 5,000 feet long and you believe there is a full-service FBO on the field, there are some other questions to ask and get answered before assuming that this is the airport for you. How are your crosswind landing skills?

Back in the mid-1960s I fell into this trap while ferrying a Cessna 180 from Ohio to Oregon to its new owner. The airport in the northern part of Colorado where I planned an intermediate fuel stop was such a place. Everything looked great until I realized while approaching the traffic pattern of this uncontrolled airport that there was a whale of a crosswind blowing directly across the runway. This particular leg of the cross-continent flight was several hundred of miles long. I left an airport earlier that morning where the wind was almost calm. As the trip progressed, I neglected to keep aware of the surface winds.

I was cruising at 10,500 feet on a westbound heading and expected the strong headwind that was in my face at altitude. Unfortunately I wasn't alert to the problem about to be encountered upon landing. Because the leg had been long and the fuel was running low but not yet critically so, I blundered into an adventure for which I wasn't mentally prepared. Being young and one of less wisdom in those days, I went ahead and flew the pattern and set up my preferred wing-low method of dealing with the final approach to a crosswind landing.

When the time came to initiate the flare, I found that the airplane had reached its aerodynamic limit of crosswind controllability. Thanks to a piece of optional equipment on that airplane known as *crosswind landing gear*, I didn't tear up the machine and was able to get it on the ground in one piece.

The taxi from the runway to the parking area was another adventure in keeping the airplane right-side up. It required a considerable amount of time taxiing with the cross-wind gear "kicked out" from the normal straight-ahead position into the position where the main wheels were castered so that the airplane was actually crabbed into the wind while rolling on its gear.

That was the last time that I failed to keep a good eye on the wind. In the future, I opted either not to make a certain trip if the winds were howling from the wrong direction at the destination or to choose a different termination airport.

The type of airplane that you fly and your pilot skills will determine just how much crosswind is too much. Your airplane's manual has a maximum demonstrated crosswind component specified, but that number doesn't tell the whole story. Too often the maximum component was all that was available during flight testing for certification of the airplane. In that case, the book number doesn't represent the limit of aerodynamic controllability. Sometimes the book number will represent the limit if the certification test pilots were able to find some very strong winds during their testing.

You have no way of finding out which is the case for the maximum demonstrated crosswind component specified for your airplane. Before you would consider doing any testing to find out, make sure that you feel good about flying in high winds, regardless of direction. Fly at an airport with a runway pointed into the wind to prevent getting caught in an untenable situation of taking the test beyond the limits.

I agree with the proposition that it's easier to land a high-wing airplane in a strong crosswind than it is to do so in a low-wing aircraft. Taxiing is another matter because most low-wing airplanes are much easier to taxi safely in high winds than are high-wing airplanes. Like everything else about aviation, there's a compromise involved and some part of landing in a strong crosswind will tax the airplane and the pilot.

High-density traffic area operations demand another kind of pilot proficiency. If you routinely fly in uncongested areas, have never flown into busy airspace, or were uncomfortable with it when you did fly in busy airspace, a cross-country flight is not a good time to fly in congested airspace without some additional instruction to deal with the unique problems encountered at high-traffic airports.

Nothing about flying ought to be learned in the self-taught environment of getting into a situation that you either haven't been trained to handle or with which you're uncomfortable. Get some dual instruction or at least go into a high-density traffic area with another pilot who is familiar with such operations.

The controllers have every right to expect that those pilots flying in their airspace know what they are doing, know how to use ATC services, and have familiarized themselves with the terminal area itself. First, get a terminal area chart for the area if one is available. Closely study the chart before you take off. Know the location of and use the flagged VFR reporting points that are depicted on the chart. Be sure that you are up-to-date on the airborne equipment requirements in the FARs that dictate what avionics must be operational in your airplane for that type of airspace. We'll delve into this area in detail in the next chapter.

If you don't fly into controlled airports very often, brush up on any communications skills that have become rusty. Even if you're flying into a satellite airport in a busy terminal area, the radio talk can come fast and furious because the controller at the primary airport is dealing with all of the aircraft flying in that area, whether departing, arriving, or transiting.

Regardless of the basic type of airport that you have chosen for your arrival point, either final or intermediate, study the sectional chart and airport directory information about it before you start down from cruising altitude. See if there are any special operational rules in effect, such as a nonstandard traffic-pattern direction or altitude.

Some airports require the use of a standard left-hand traffic pattern for one runway, and a right-hand pattern if landing in the opposite direction on the same strip of pavement. This occurs when there are obstacles or noise-sensitive areas to one side of the runway that ought to be avoided by aircraft flying over the area. See if the airport has any unusual operations being conducted, such as parachuting, ultralight aircraft, gliders, or helicopters. If so, know the right-of-way rules if a traffic conflict arises between you and them.

The airport where I fly gliders is a field with two long, paved runways, and the tow planes and gliders operate from the grass beside the normal active runway. It's interesting to see how pilots of power planes react when they see that glider operations are going on beside where they thought that they had the exclusive right to land. The local airplane pilots fit right into the flow; transients often do last-minute go-arounds when they see a towplane start its takeoff roll with a glider at the end of a 200-foot rope. If the pilots had looked at the sectional chart before planning a flight to this airport, they would have seen the symbol on the chart that indicates glider operations.

Another unconventional situation that you might encounter is the presence of helicopters operating from an airport. Even when the airplane traffic pattern is the standard, left-hand pattern, most helicopter pilots will use right-hand traffic to set up an approach. This is done to enable the helicopter pilot to see whatever airplane traffic might be in the area.

It's much harder for an airplane pilot to see a chopper than it is for the helicopter to see the airplane because the fuselage of the helicopter is all that is readily apparent when it's in flight, and the rotor blades are virtually invisible when whirling at normal RPM. If you see a whirlybird coming into right-hand traffic, don't assume that its pilot is in error. From that vantage point in the helicopter cockpit, the chopper pilot is much more likely to see you in an airplane than vice versa.

## Uncontrolled airports

Unless the majority of your flying is between major metropolitan areas, you are probably going to have to land at an uncontrolled airport at some point in a cross-country flight. Many pilots prefer uncontrolled fields even if a controlled one is handy. The uncontrolled airport offers a level of time savings and other conveniences that might outweigh whatever other benefits are attendant to landing at a controlled facility.

Uncontrolled airports are indicated on the chart by the use of an airport symbol that is magenta in color. Controlled airports have blue symbols. When a control tower is not in operation 24 hours per day, the chart symbol is still blue in color. So don't assume that all controlled airports have 24-hour towers; many controlled airports that serve only general aviation aircraft are part-time towers. An airport reverts to an uncontrolled status when the tower is closed.

There are some telling facts about the midair collision hazards around airports of both types—controlled and noncontrolled. Most midairs occur during daylight VFR weather within 5 miles of an uncontrolled airport, usually when both aircraft are below 3,000 feet above ground level (AGL). Only 5 percent of the collisions are head-on. Nearly one-half of all midair collisions happen on final approach when one aircraft overtakes another and never sees it.

Some authorities blame these freak accidents on the problem that the pilot of a low-wing airplane has in seeing another airplane directly in front and below him. A high-wing airplane presents its pilot with a blind spot above and to the rear. So if a faster low-wing airplane is on final approach behind a high-wing, and the low-wing pilot is unaware of the traffic in front, the elements for disaster are certainly present.

Controllers do their jobs quite well. Most such accidents occur at uncontrolled fields; therefore, pilots should be especially vigilant when operating at uncontrolled airports. For the safest possible arrivals at uncontrolled airports, follow these practices:

**1.** Check an airport directory before taking off for any airport. Controlled and uncontrolled airports are in the *Airport/Facility Directory* (A/FD), *Airguide Publications Flight Guide,* and *AOPA's Aviation USA*. You need to ascertain the common traffic advisory frequency (CTAF). Does the field have an FSS or a unicom? Should the multicom frequency be used? What are the hours of operation of the FSS or the unicom?

When you arrive in the airport area, will you be able to talk to someone on the ground for traffic advisories, or must you rely upon calls made in the blind to other pilots? All of this information should be gleaned from one of the directories before you get there.

Traffic pattern altitudes are not given on aeronautical charts, and there is no standard pattern altitude at all airports. Pattern altitudes can vary from 600 feet to 1,500 feet AGL for propeller aircraft to as high as 2,500 feet for jets. You need to know what is expected before you arrive.

Will the FBO be open for business when you arrive, or are you on your own to find the transient parking ramp and secure your airplane? If you'll land after sunset, are the runway lights on all night, or is there an automated system involving a certain number of microphone clicks on a certain frequency to activate the lights? That can be a real surprise if you haven't checked ahead of time.

Know whether the runway length is sufficient for a safe landing and subsequent takeoff in your airplane. Realize that most airplanes can be landed in less distance than that required for a safe takeoff, especially considering that during a cross-country flight most landings are made at a reduced aircraft gross weight; the next takeoff will probably involve higher weight and therefore require more runway than landing. Check NOTAMs for any runway and taxiway closures, repairs, or nonstandard hazards.

**2.** When you are about 10 to 15 minutes out from the airport, but no farther than 15 miles away, tune your radio to the CTAF and listen. Try to determine the active runway in use and the general nature and level of airport activity.

A good anticollision technique is to turn on your landing light before entering the traffic pattern during the day as well as at night. Landing lights are great anticollision lights and can be seen at great distances if the angle of the viewer is right. It should go without saying that the airplane's other normal anticollision lights should be on, whether strobes, a rotating beacon, or both. Remember that you're in "see-and-avoid" territory.

**3.** Check the airport diagram well in advance of arrival, and visualize the runway in use and the traffic pattern around it. Broadcast your intentions and position when you're about 10 miles out. If the unicom is in operation, you might hear a response, but don't count on it. If you get a reply, ask for the runway in use, winds, and other traffic in the area if that information is not volunteered by the unicom operator.

Remember to listen as well as transmit. I fly gliders at an uncontrolled field that has two long runways and serves airplane traffic at the same time that our glider club is flying. You'd be surprised how many pilots announce their positions and intentions on the CTAF frequency, but never acknowledge a response from our tow planes and gliders nor seem to have understood the responses made to their calls.

Your job isn't just to let other traffic know what you intend to do; your job is also to find out what the rest of the traffic using that airport is doing. If no reply is forthcoming, you're in the blind. Still broadcast your entry into the pattern, and definitely announce your presence on the final approach; remember that's where most collisions occur.

If you have any lingering doubts about any aspects of the airport (active runway, wind direction and velocity, amount of traffic, etc.), there is nothing wrong with overflying the field if you do so at least 1,000 feet above the traffic-pattern altitude (always at least 2,000 feet above the airport elevation) and check out the details.

If you engage in this practice, one note of caution is in order. That is to be sure that you distance yourself away from the airport enough to be able to descend to traffic pattern altitude before you actually enter the pattern. It's very dangerous to enter the traffic pattern while descending, especially so if you fly a low-wing airplane. Always do your best to plan to be at the pattern altitude before entering the pattern. A good rule of thumb is to be a minimum of 2 miles out from the airport at the time that you reach the pattern altitude.

**4.** Announce your type of pattern entry on the CTAF. Always preface your call with the airport name, then give your type aircraft (sometimes even saying the color of the airplane helps), call sign, your position, and lastly your intentions. Remember that other pilots in the area can't see your N numbers. Type of aircraft is far more important to an "in the blind" call than is the call sign.

Four Ws are always the key to any radio call:

- Whom you are calling
- Who you are

- Where you are
- What you want to do

Remember these, and you'll only have to modify the actual words of your calls slightly to fit the circumstances that confront you. A typical blind CTAF call goes like this: "Delaware traffic, yellow Cessna 4517 Charlie, 2 miles south of the field to enter downwind for Runway 28." That call gives every other pilot in the vicinity the information needed to avoid a conflict with you.

Even though straight-in approaches are no longer illegal by the Federal Aviation Regulations, there is seldom a reason to use one. Straight-ins deprive you and whoever else is operating near the airport of all-important time and opportunities to visually acquire any conflicting traffic. Using a straight-in is a very good way to either overtake another airplane on final, or be overtaken yourself, which is the primary scenario for a midair collision. Fly a standard pattern.

**5.** After making your CTAF call, keep your head on a swivel for other traffic, and enter the downwind leg. Once established on downwind, make another call (remember the four Ws) and advise traffic of your intentions, whether you're landing full stop, doing a touch-and-go, or a low approach.

Whether to make additional calls as you turn base and final is a judgment call. The locals call every leg at some airports, but that might not be the case at other airports. The idea is to give everyone the safety margin of knowing what else is going on around them, while at the same time not being overly verbose or clogging the frequency with idle or unnecessary chatter. Follow the old adage of "In Rome, do as the Romans do," and you'll fit right in.

**6.** Keep the lengths of your pattern legs reasonable. If you are used to flying at large airline airports, don't make a 2-mile downwind at an uncontrolled airport. That's asking to get cut out by the airplane following you or another pilot who flies short pattern legs.

Watch the other traffic ahead, and fit into their pattern. Everything works a lot better if the pattern remains predictable. Naturally if there are more than one or two other airplanes in the pattern with you, leg lengths will have to stretch out a little; just be reasonable and flexible.

**7.** Get off of the runway as soon as safe and practical after you land. If there is no taxiway, you'll have no option but to do a 180-degree turn on the ground and taxi back up the runway upon which you've just landed. As soon as you can do this, the sooner that you will be able to see if there are any airplanes about ready to land. Perhaps any landing pilot will be able to let you get off the runway prior to his touchdown. If there is a taxiway available, clear the runway and get on that taxiway as soon as it is safe to do so.

Pilots who regularly fly into such airports expect you to be considerate and not cause unnecessary go-arounds by hogging the runway for an exorbitant amount of time. I've seen airplanes land on these airports and slowly taxi for thousands of feet on the only runway, hogging it without a thought of what might be happening or about to happen behind them.

Again, being reasonable is the answer because no one expects you to squeal the brakes and wear a flat spot on the tires with excessive braking, nor are you expected to turn off the runway at high speed. Use common sense and courtesy.

Be wary of taxiing on grass adjacent to the runway. It might be safe; it might be unsafe. These areas might contain storm sewer grates, chuck holes, or other similar hazards. If you don't know the airport very well, stay on the pavement.

When leaving an uncontrolled airport, keep these thoughts in mind:

**1.** Make a radio call on the CTAF as you taxi to the runway. If the unicom is monitored, you should get the typical wind, runway, and traffic information. If not, at least you're letting other pilots in the pattern know that someone will soon be at the departure end of the runway.

**2.** Occasionally check the traffic pattern while you taxi, and pay attention to the final approach leg. If the wind is calm, another airplane could be landing on any runway, so don't assume that because you don't see anyone coming to the runway that you will take off from that the entire airport has no other traffic.

Often in the evening at rural airports that have a runway that faces into the setting sun, you'll find that local pilots might land the other way, even if that involves a slight tailwind on their final. If you're taking off into a low sun, your visibility will be reduced, so be certain that someone else isn't headed into your departure path.

**3.** When you are ready to take off, make a final visual scan of the traffic patterns of all of the runways at the airport. When you are satisfied that the entire area is clear, pull out onto the departure runway and take off promptly. Have all of your pretakeoff checks and chores done before you taxi into takeoff position. Don't sit on the runway with your back to the traffic pattern.

The whole story of safe operations into and out of uncontrolled airports is to remember that it is every pilot's duty—your's and mine—to see and avoid other traffic. Plain old common sense and the sensitivity to what others might be doing will go a long way toward accomplishing that end.

## Controlled airports

We'll have a whole lot more to say about operating at controlled airports in chapter 3 because most controlled airports lie within either Class B or Class C airspace. But there are a significant number of controlled airports that don't lie within either of these classes of airspace and are within Class D airspace.

This simply means that at controlled airports in Class D airspace you probably won't be talking to any radar controllers before communicating directly with the controller in the tower who is exercising jurisdiction over the aircraft in the traffic pattern or departing from the field.

Class D airspace is the new term for what we formerly called control zones. In September 1993, the United States got on board with the rest of the international aviation community, and redesignated its airspace classifications to conform to the worldwide scheme.

The airspace within a Class D area is cylindrical in shape, centered around the controlled airport. The top of the cylinder generally goes up to about 2,500 feet AGL. No longer is this upper limit of the designated airspace uniform as it was before September 1993. The sectional chart shows the ceiling of each piece of Class D airspace, so look for it.

If you plan to overfly a Class D area, that's fine as long as you're above the ceiling of the controlled area, the Class D cylinder of airspace. The ceiling is shown on the chart so you can easily determine the minimum altitude to fly and remain above the cylinder. Common sense would tell us that it doesn't make much sense to press the issue and fly as close as possible to the ceiling of the Class D area.

If weather conditions permit, give yourself some slack and stay at least 1,000 feet or more above the published ceiling of the controlled area. You'll accomplish two objectives. First, you'll be that much farther away from the traffic operating within the Class D; therefore, you will be that much safer. Second, the tower controllers will more easily realize that you are an aircraft outside of their jurisdiction and won't be concerned with wondering what your intentions are; otherwise the controllers might think that you are inside their airspace without communicating and try to file a regulatory violation against you.

The Class D cylinder is generally about 10 miles in diameter, 5 miles in all directions from the airport center, but this can vary slightly. When you're around 15 miles or so from the airport, you should be tuned into the control tower frequency. If the airport has an automated terminal information service (ATIS), you should have first listened to it.

When you listen to the ATIS and digest that information first, the rest of the process of dealing with any controlled airport will be a lot smoother. One of the biggest problems at all controlled fields is frequency congestion. Two-way radio is inherently limited to one-way transmissions; no one else can talk when a pilot or controller is transmitting.

When you get into a beehive of traffic, many pilots must talk to one controller, and the solitary controller has to talk back to every pilot. Procedures have to be in place to reduce the number of words that everybody says. This ultimately reduces the time that the frequency is tied up by the pilots and controller. By using the ATIS, you'll already know the basic weather conditions, the runway(s) in use, the winds, and any special instructions germane to all pilots approaching the airport.

The ATIS recording is changed every hour, and each time it is revised the recorded voice will give you the identifying phonetic letter for that hour's information. After you've listened to the ATIS, remember the phonetic letter attached to the recording you heard. When you make your first call to the tower, let the controller know that you've already listen to the ATIS.

A call could go this way, "Ohio State Tower, Cessna 4517 Charlie, 15 north, landing with Information Bravo." You've obeyed the 4 Ws of radio communication because you addressed to whom you're talking, said who you are, and told the controller where you are and what it is that you want to do. During your first call to any ATC facility, use your aircraft type and full call sign, but you can omit the "November," which is the phonetic alphabet word for the letter N, the first letter in the registration numbers of all aircraft of United States registry.

After the first call, you can refer to your airplane by the last three numbers or letters of its call sign (phonetic words for letters, please). Your first call would provide all the data that the controller needs to fit you into the flow of traffic approaching the airport.

If the airport isn't too busy, the controller might respond, "Cessna 17 Charlie, enter right downwind for 27 Left, report abeam the tower." The controller's response took only a few seconds. Two short and efficient transmissions by a conscientious pilot and attentive controller got the job done with minimal frequency use.

If you hadn't listened to the ATIS, the controller would have had to tie up additional time giving you the winds and altimeter setting, and your mind wouldn't have been primed with advance knowledge of the runway in use, and what wind conditions you would have encountered when landing.

Another of the rules of two-way radio communication that applies universally to normal use is to always tune your radio to the appropriate frequency and listen for a few seconds before transmitting. Only one person can talk at a time, and if you jump in without listening first and transmit while someone else is doing so, all you'll accomplish is to give everyone on the channel a loud squeal in either their headsets or speakers. By listening first, even on a busy frequency, you can pick an opportune moment to interject your transmission to the flow and keep everything going much more smoothly.

After you've entered the downwind leg of the traffic pattern at our hypothetical example of Ohio State University Airport, the tower's instructions were to report abeam the tower. This means that you are to enter the downwind in a normal fashion, about midway along the length of the runway, and fly until the control tower is off of your right wingtip (because you're on right downwind).

Then simply transmit "Ohio State Tower, 17 Charlie, abeam the tower." Chances are good that you'll then hear back "17 Charlie, cleared to land." That means that you're cleared to fly the remaining portions of the pattern and execute a normal landing on the previously assigned runway, which was 27 Left.

If there are other airplanes in the traffic pattern, you'll be advised and instructed to follow a certain airplane ahead; the controller will tell you what type of airplane it is. If you don't have that traffic in sight, tell the controller "Negative contact"; otherwise the controller will assume that you do have visual contact with the aircraft that you're supposed to follow and you will maintain proper separation.

When you don't see your traffic ahead and you have properly reported that to the controller, it's up to the tower to keep you separated. If you have previously reported the traffic in sight, and then lose contact with it, tell that to the controller too. Even so, continue to scan the pattern ahead and try to pick up the preceding traffic. Always watch for unreported traffic, too.

When you're following other traffic at a controlled or uncontrolled airport, keep a reasonable but not excessive distance between the two of you. Extra-long pattern legs only bottle things up more for the airplanes behind you. At a controlled airport, you should know what their intentions are based upon the calls from other pilots; this also applies to an uncontrolled field if the traffic is making CTAF announcements.

If the airplane ahead is making a touch-and-go, you could naturally follow a little bit closer than if the airplane were making a full-stop landing because the landing airplane does not need the extra time to slow down and safely taxi off the runway after landing.

When you're following a large airplane ahead of you, you'd better know how to avoid the wake turbulence generated by it. Even with all of the best knowledge and intentions that pilots can have about wake turbulence avoidance, accidents still happen when small airplanes encounter the wakes of large airplanes.

Avoiding wake turbulence should be a shared responsibility. The controller should always warn you about its potential and not give you instructions that might put you in peril. Every pilot should take responsibility for actions in the cockpit. Always allow plenty of room and time for the vortices to move out of the way and dissipate.

After your landing roll, you'll probably be told to contact ground control on the appropriate frequency for that airport. Don't switch off the tower frequency until you've actually left the active runway. Remember that a clearance to taxi to the parking ramp implies a clearance to cross any inactive runway that you might encounter en route.

Collisions have occurred on inactive runways, so keep alert and look for traffic on any runway that you have to cross. Controllers are human and make mistakes like the rest of us; pilots can inadvertently take off or land on the wrong runway. Thankfully those errors are few and far between, but no one wants to be the subject of those that do seldom happen.

If you're arriving at a strange airport, especially at night, you might not have a clue as to where you want to go on the airport, or how to get there. Only the proud feign familiarity with the area; the wise ask for a progressive taxi. That means that you need the controller to lead you from wherever you are to your destination. It makes things a lot simpler to get a progressive than to end up in some strange corner of the airport, or, worse yet, to wander onto an active runway.

When you're ready to depart from a controlled airport, the first thing to do after getting the airplane preflighted and started is to once more tune into the ATIS and listen for the recorded information. When you call ground control for the initial taxi, tell the controller that you've got the information (by the phonetic hourly identifier).

Quite a few of the busier controlled airports, especially those within either Class B or C airspace, have a clearance delivery frequency that is separate from ground control. The purpose behind the inception of clearance delivery channels is to free the ground controller from the job of procuring and reading IFR clearances to outbound IFR airplanes.

Without this task, the ground controller can devote all attention to keeping airplanes apart, which is his primary duty. A distinct clearance delivery frequency allows pilots to call before engine start to get their clearances. If there is a delay in getting cleared, airplanes aren't waiting needlessly at the end of the runway or unnecessarily burning fuel.

Because we're concerned with VFR operations, you must keep in mind that where they have been established you too need to call clearance delivery before you call the ground controller. Still, listen to the ATIS first and give the clearance delivery controller the phonetic identification of the recorded information that you just heard.

When making your call, it can go something like this, "Columbus Clearance, Cessna 4517 Charlie at Lane Aviation, VFR to Dayton, with Information Papa." The controller will respond, telling you that you're cleared out of the Class C airspace around Columbus, Ohio, and will further give you a maximum altitude to observe while within the Class C, a transponder code to squawk, and the ground control frequency.

Then, when you call ground control, all that is needed is, "Columbus Ground, Cessna 4517 Charlie, at Lane, ready to taxi." The ground controller can then limit her instructions to telling you which runway to taxi to, and any special instructions en route, such as the need to hold short of any intervening inactive runways, or taxiways to give way to other aircraft. Everyone talks less and the traffic gets moved more safely.

When you're at the departure end of the runway, don't change frequencies off of ground control until after you've performed the preflight runup and checklist. When you do change to the tower, you should be ready to go. When cleared for takeoff, give the final approach course to that runway the same visual scan that you would at an uncontrolled field. Too often somebody lands on top of a departing airplane, and a surprisingly large number of these mishaps occur at controlled airports.

At most airports within Class D airspace, you probably won't have to deal with a radar departure controller. Normally when you call the tower ready for takeoff, just tell the controller what kind of departure you want (straight-out, left turn, or right turn), or tell him the direction in which you want to leave the area. Unless something unusual is happening at the airport, or the traffic is heavy, you'll get your request approved and be on your way.

Passing the outer rim of the Class D cylinder during the climbout, some pilots prefer to call the tower to say that they are leaving the controlled area. Other pilots monitor the tower frequency for several miles, and then leave that channel when well outside of the tower's jurisdiction. There is really no right or wrong way to do it. Frequency congestion is the best guide; if it's busy that day, there is nothing to gain by tying up the radio to say goodbye; if it's a slow day, you might want to.

If you aren't too current on communications skills, one way to sweep out the mustiness is to listen to the frequencies of a controlled airport while not operating close to it. That can be done by listening in to a nearby controlled airport while flying outside of its airspace. Most towers can be heard at least 40 or more miles out if you're flying high enough that the line-of-sight limitations of VHF radio don't prevent hearing the transmitter. The airplanes in the controlled area are naturally quite a bit higher than the tower, and their broadcasts can sometimes be heard for nearly 100 miles.

If you live in a metropolitan area with a controlled airport, think about getting a hand-held communications radio to use to listen to the airport's radio chatter. Every pilot who does much cross-country flying ought to have a hand-held radio anyway, if for no other purpose than as a source of communications if the airplane's radios or electrical system go on the fritz. They also make great training aids because you can listen from the comfort of your family room chair.

Regardless of the type of airport that you use, each presents its own challenges to be met and mastered. For every knowledgeable and prepared pilot, that's part of the fun of cross-country flying.

# 3
# Airspace classifications

ALL BUT THE SHORTEST AND MOST ELEMENTARY CROSS-COUNTRY FLIGHT will take a pilot through several different classifications of airspace. Each type has differing rules applicable to operating in it, and every pilot is expected to know the regulations and comply with them.

Most of the FAA violation cases pressed against private pilots concern allegations of failing to obey the rules about flying in various classes of airspace. The system can appear intimidating at first, but it isn't really all that hard to understand. Don't give up when you take your initial glance at Fig. 3-1 because we'll make some sense of it together as we go.

## AN ENTIRELY NEW SYSTEM

In September 1993, the FAA tossed out the airspace classification scheme with which almost all then-licensed pilots grew up and used for decades. In this current world of global economies and rapidly changing commercial and political alliances, our outdated scheme had to go.

What we have now is a system of airspace classification that concurs with that in use throughout most of the world. Each country has some variations in the specifics, but the United States no longer has a system that made no sense to pilots from any

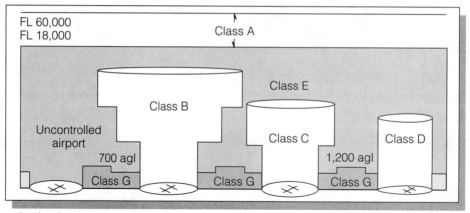

*msl - above mean sea level
agl - above ground level

**Fig. 3-1.** *The airspace system in the United States.*

other country. If you're a newly certificated pilot, you're lucky because you didn't have to unlearn one system and try to assimilate a new one into your brain. For the rest of us, it's a necessary burden, albeit an unpleasant one.

Rather than using names or words to identify the various classes of airspace, as was the prior practice, the present scheme uses letter designators. They are Classes A, B, C, D, E, and G. There is no Class F in the United States, although that identifying letter is used in other parts of the world.

## Controlled and uncontrolled airspace

The first premise from which we should start our discussion is to define what it means to say that some airspace is controlled, and some airspace is uncontrolled. This is simple. All that it means to say is that controlled airspace is any airspace where flights can occur under the control of ATC. Uncontrolled airspace is where flights under ATC control do not operate.

Don't be confused—laymen and student pilots of this and past generations have been—that you have to be under any ATC control to fly in controlled airspace. Most flight in controlled airspace is done under VFR without talking to ATC or being controlled by anything other than the pilot's wishes. All that controlled airspace means is that if you are flying under IFR, you've got to remain within controlled airspace, except during transitions and approaches, because ATC controls IFR operations.

Certain types of controlled airspace, particularly Classes A and B, require the pilot to be under the control of an ATC facility when within them. Classes C and D require communication with a controller. Class E, which can be characterized as general controlled airspace, does not require talking to anyone unless you want to. Just remember that because piston-powered IFR traffic generally operates en route within Class E, it is described as controlled airspace because those IFR flights are being controlled by ATC.

Class G is called uncontrolled airspace because it comprises the areas where there is no control of traffic (no en route IFR operations). For too long pilots have been confusing this subtle distinction. Many pilots never understood the old system and are immediately lost when trying to make sense of the new one.

## Class A airspace

As a pilot of a typical single-engine or light-twin piston-powered airplane, chances are quite good that you might never venture into Class A airspace. But with the increasing popularity of turbocharged engines that are very capable of operating in the flight levels, some general aviation pilots do fly high enough to get into Class A.

Class A airspace is all of the airspace in the United States above 18,000 feet, or as more commonly defined, above Flight Level 180. When you fly at or below 17,999 feet, all pilots are expected to use normal altimeter settings based upon the corrected barometric pressure of the air mass in which they are flying. At 18,000 feet, all altimeters are set at standard barometric pressure, which is 29.92 inches of mercury. When flying that high, terrain clearance ceases to be an issue in all but the most mountainous areas of the West, and even in that region few mountains rise high enough to concern a pilot flying at or above 18,000.

Also, most airplanes that operate in Class A airspace are turbine-powered, and naturally have much faster cruise speeds than do the aircraft that fly lower and slower, according to visual flight rules. Pilots of these high-speed aircraft would be constantly changing altimeter settings as they zoomed over the reporting stations.

This problem is eliminated by the convention that has been established for all aircraft operating in Class A airspace to have a common altimeter setting of 29.92. Because everybody's altimeter is set to the same value, you get the same vertical separation capability as if the altimeters were set to the actual pressure then in effect.

When a pilot has climbed to 18,000 and resets the altimeter to 29.92, she might have to continue to climb in order to reach FL 180. If the local station pressure in that locale is such that FL 180 would require a descent from 18,000 feet MSL, that flight level will not be used, and the lowest usable flight level in that area on that day would be FL 190.

Flight levels are designated by a three-numeral system that simply drops the last two zeros from the corresponding height if it were measured in feet above sea level. So 18,000 feet becomes FL 180; 20,000 feet is referred to as FL 200, and so on up to FL 600, which is the upper limit of Class A airspace.

No normal civilian traffic is allowed above FL 600, and the only flights conducted above FL 600 are in military or other government or specially authorized aircraft.

Operations within Class A airspace are restricted to only those flights operating on an IFR flight plan under the positive control of ATC. VFR is strictly prohibited. Because of the IFR requirement, the pilot flying in Class A must possess an instrument rating and must also meet all of the instrument currency requirements contained in FAR Part 61. This mandate must be obeyed regardless of the actual weather conditions.

For some reason there has always been some confusion among the general aviation pilot population about whether you must be IFR current to file an IFR flight plan, accept an IFR clearance, and operate under ATC control when the weather is clear as a bell and there is no chance of encountering real IFR weather. Dispel any misconceptions that you might have. The regulations are very clear on this point; you must be instrument-rated and current to operate under instrument flight rules, regardless of the actual weather conditions at the time.

If you've lost your IFR currency, you must regain it one of two ways. Within the six-month grace period, fly the required six hours of instrument time and six approaches; outside of that grace period, your only remaining option is to get an instrument competency check from an instrument instructor, or its equivalent, as set forth in FAR 61.

Because there is no VFR flight permitted in Class A, there are no VFR visibility or distance-from-cloud minimums stated in the regulations. As an aside, when flights in Class A are conducted above Flight Level 290, ATC will increase the vertical separation between aircraft to 2,000 feet instead of the 1,000 feet in effect at lower altitudes. This increase is used in light of the fast speeds of airplanes that fly that high.

The regulations also require that aircraft flying in Class A airspace have to be equipped with certain avionics, including a Mode C transponder with altitude encoder, VOR, and DME. Naturally, because of the altitudes covered by Class A, the aircraft must either be pressurized or equipped with supplemental oxygen for all occupants. If the flight is going high enough, you are required to have oxygen available, even if the airplane is pressurized, so that an unexpected loss of pressurization can be handled and the crew and passengers can breathe during the time that it will take to descend to a lower altitude.

## Class B airspace

Class B airspace is also positively controlled by ATC. This class exists around high-density air-carrier airports in most of the major metropolitan cities in the nation. (Class B was formerly referred to as a terminal control area, or TCA.) The shape of a Class B area of airspace is often analogized to an upside-down wedding cake (Fig. 3-1).

Class B airspace is shown on sectional charts by a series of solid blue lines that make concentric circles around the primary airport. These lines identify the tiers of the wedding cake, and inside each segment there are two numbers, one on top of the other, separated by a horizontal blue line.

If you see in one of the segments the number 80 over the line and 40 beneath the line, that means that in that layer of the Class B airspace, the ceiling of the controlled area is 8,000 feet, while the floor of the controlled area is 4,000 feet. In the same way that flight levels are depicted, the last two zeros of the altitudes in question are deleted.

Both IFR and VFR flights can operate within Class B airspace, and there is no requirement for the pilot to be either instrument-rated or IFR-current; however, at most Class B locations, the pilot must be at least a private pilot.

Student pilots are allowed to fly in a few Class Bs, but to do so they must have had instruction both on the ground and in flight from a flight instructor in operations in the specific Class B airspace for which the instructor is going to authorize solo flight. The instructor must endorse the student's pilot logbook to reflect the required instruction and to certify that the student has been found competent to conduct solo flights within that specific Class B area. This endorsement must be renewed every 90 days.

Recreational pilots are not permitted to fly as pilot in command in any airspace that requires communication with ATC, so that prohibition extends to Class B airspace.

The ceiling of the Class B airspace will be set at about 8,000 feet above ground level AGL, but this will vary according to local needs of the airport at the center of the "cake." The Class B airspace will normally extend out from the airport in all directions for 20 nautical miles, giving it a diameter of 40 miles; the size can be modified to meet local requirements.

The first tier of controlled airspace is at the center of the Class B airspace and extends from the surface of the ground up to the upper limit of the Class B. This segment of the scheme is usually about 5 nautical miles in radius from the center of the airport. Everyone operating within this 10-mile-wide cylinder must be under a clearance issued by ATC and talking to the controller. Occasionally you'll see notches cut out of this part of a Class B in order to accommodate satellite airports.

Working from the center out, the next layer of the upside down cake will have a floor. Beneath that floor pilots can operate without actually being in the Class B and are therefore relieved of the requirement to be on a clearance and flying under the control of ATC. This second tier will normally start at the lateral boundary of the cylinder that extends from the surface up to the ceiling of the Class B, which is again generally about 5 or so miles out from the primary airport in all directions.

The second layer usually has a lateral coverage area of about 3 miles. What you'll see in the second tier is that the floor is quite low, often about 1,100 feet or so above the surface. Within this area, pilots may fly without being in the Class B itself, but at an obviously lower altitude than is generally the norm for cross-country flights. The purpose of this layer is to allow for operations at and around satellite airports.

Because Class B airspace exists around major cities, there are often urban obstacles to flight, such as antenna farms, large industrial stacks, and similar hazards within the second layer. If you are circumventing a Class B area on a cross-country trip, it's best to do so at an altitude that is at least underneath the third layer of the upside-down cake. It's too congested and hazardous to fly low enough over major cities to stay below the floor of the second layer.

The third tier of Class B airspace will begin at the outer edge of the second layer and extend for about another 7 nautical miles in an ever bigger circle of coverage. Now the floor of the Class B will rise to about 2,000 feet above the surface, perhaps slightly higher. *In this area below the published floor as shown on the chart, you can fly with-*

*out a clearance or calling the ATC facility exercising jurisdiction over the Class B airspace.* Because the floor has now risen to a reasonable altitude above the ground, it is often possible to use this distance from the center airport when flying cross-country and going around a Class B.

The last ring of a Class B will extend from the third layer out about 5 nautical miles. Here the floor usually is set at around 3,000 feet above the ground. Things can still get busy this far out because experience has shown that most pilots try to circumvent a Class B with as little detouring as is consistent with safe flying; therefore, many pilots choose to do their Class B avoidance within the outer ring, staying beneath its floor.

The government produces a terminal area chart for each terminal area that is Class B airspace. These charts show the same information as do sectionals, but everything on them is blown up to a scale that halves the geographical coverage of the chart, compared to the area shown on a sectional. Before you operate in Class B airspace, you should purchase the applicable terminal area chart. An area chart makes visual flying much easier, and you can more readily ascertain the landmarks that you need to use to stay within the various segments of Class B or completely avoid the area if that is your desire.

Remember that Class B airspace is an extremely congested place; that's why the Class B exists. If you intend to go around it without talking to a controller, give the center airport as much berth as practical, and keep in mind that dozens of other pilots are doing the exact same thing.

There is nothing to prevent a pilot who is at least a private pilot from contacting the ATC facility governing the Class B airspace before entering it and asking for a clearance through the Class B; the controller will want to know the intended route and altitude. Just look on the chart for the appropriate frequency to contact ATC; the frequency will depend upon the aircraft's bearing from the airport.

The location and the present levels of traffic will have a bearing on whether the request can be granted in full, refused, or granted with instructions to fly at either a different heading or at a different altitude or both. Some of the airports with Class B airspace around them, such as Cleveland, Ohio, do not have the traffic levels that are prevalent at places like Atlanta and O'Hare.

When traffic permits, controllers will generally be accommodating to a general aviation pilot's request to traverse their airspace. Use common sense, and if your request is denied and you are not issued a clearance to enter the Class B airspace, be understanding and comply.

Many people are confused about what constitutes a clearance to enter Class B airspace. They shouldn't be. When the controller says "Cessna 4517 Charlie, cleared to enter the Class B airspace maintain VFR at 2,500, fly heading 030," you are then permitted to enter. Without the magic words "cleared to enter," you are not permitted in. The confusion arises because most noninstrument-rated pilots are not accustomed to operating on clearances and don't understand the precise meaning of that word in aviation parlance.

A clearance to enter the Class B airspace will contain three items:

• Permission to enter
• An altitude to maintain until told otherwise
• A heading or route to fly

You are obligated to follow the directions precisely. If you're not on an IFR clearance, be careful that the instructions from the controller don't take you into IFR conditions. If that possibility exists, tell the controller at once that you cannot maintain VFR if you do as told, and ask for a different altitude or heading.

The VFR minimums within Class B airspace have changed from what they were under TCAs. You still have to have 3 statute miles of visibility, but you are only required to remain clear of clouds with no minimum distance from them. That makes sense because all of the traffic in Class B is under the control of ATC, and traffic should not unexpectedly pop out of the clouds to create a hazard; ATC is providing the separation.

The controller will provide traffic alerts if there is another airplane that presents a potential conflict to your flight. Again, if you think there's a problem coming, tell the controller before you have to deviate from the clearance. Also remember that you must remain VFR unless you're instrument-rated and flying an instrument-equipped aircraft. If you can't raise the controller in time to get an amended clearance before entering IFR conditions, do what you need to do to remain VFR, but tell ATC as soon as possible.

One other point needs to be made about operations in and around Class B airspace. The FARs require the aircraft to be equipped with both a two-way radio and an altitude-encoding (Mode C) transponder. As far as the radio goes, make sure that you have a 720-channel transceiver before venturing into Class B areas. There aren't too many of the older variety radios around any more that had only 360 communication channels; antiques with just 20 or so crystals are essentially nonexistent. If you aren't able to talk with the controller on whatever frequency is assigned, you've got a major problem and maybe a regulatory violation on your hands.

The transponder has got to be one with the full 4,096 codes available and must have altitude-encoding capability, which must be used. In addition to requiring the transponder for flights within Class B airspace, the regs now also mandate that *any* flights within 30 nautical miles of the Class B airport must have an altitude encoding transponder in the aircraft, and must be operational with encoding enabled. This requirement is what is known as the *Mode-C veil*. A large blue circle will be depicted on the chart, showing the 30-mile radius of the center airport within which the transponder must be used on Mode C.

After a lot of discussion when the Mode-C veil was first proposed, an exemption was made for those aircraft that were not originally certified with an engine-driven electrical system or subsequently modified to include one. These older classics, such as the Aeronca Champion and Taylorcraft, can fly inside the 30-mile Mode-C veil but cannot enter Class B airspace. The veil extends from the surface up to 10,000 feet MSL or the ceiling of the Class B airspace, whichever is lower.

If your airplane has a transponder and it is not functional, FAR 91 allows you to ask ATC for a deviation from the requirement to have an operational transponder. The request may be made over the radio when you contact the ATC facility; make sure that you do it before entering the 30-mile veil. There might be times when weather and traffic considerations will dictate that the controller won't be able to honor your wishes, so be certain that you have an alternate plan in mind in case your request has to be declined.

It is legal but not necessarily wise to fly over Class B airspace. All Class B airspace areas have a ceiling, and above that ceiling you can motor right over the primary airport. When you do so, you're on your own to see and avoid the large volume of airline traffic that justified the creation of the Class B.

Also keep in mind that the airspeed limit of 250 knots only extends up to 10,000 feet MSL for most purposes. When the airliners and other high-performance airplanes climb through 10,000 feet, they routinely start accelerating right away. Most of us who are flying VFR don't want to be staring down the barrel of a 747 doing 300 or 400 knots.

## Class C airspace

This class of airspace under the new system inaugurated in 1993 was formerly called an airport radar service area (ARSA). These were primarily established around airline airports that did not have enough traffic to justify onerous restrictions inherent in high-density areas. Class Cs are very similar to Class Bs, but the rules and methods are somewhat simplified.

An area of Class C airspace will have only two layers of application. They are shown on sectional charts as solid magenta concentric circles around the primary airport. The first circle, just like in Class B, will extend from the surface up to the ceiling of the Class C airspace. The ceilings of Class Cs are substantially lower than Class B ceilings, usually going up to about 4,000 feet AGL.

The inner segment of the area will generally extend out from the airport for 5 nautical miles, while the second tier will go from that 5-mile point out another 5 miles, for a total of 10 miles in all directions from the center of the primary airport.

The second outer layer usually has a floor around 1,200 to 1,700 feet AGL. Sometimes the floor of the outer layer might change throughout its arc, so be certain that you study the chart before making any assumptions or quick judgments.

The primary difference between flying in Class C airspace as compared to Class B is that in Class C there is no requirement for a clearance from ATC before entering. All that the regs require is that you establish two-way radio communications before going into Class C and maintain such while within it. Communications are not established until you call the controller and *a reply is received*. Your making the call doesn't establish communications, so be sure to institute the process well outside of the lateral boundaries of the Class C area to allow enough time for your call to be *answered* before you enter.

If the controller doesn't want you to enter the Class C airspace, you'll be promptly told. If you get such an instruction, you must adhere to it. Although you're not technically on a clearance, you have to follow what the controller tells you to do because she has the duty to keep all aircraft inside the Class C airspace separated from each other.

Circumventing Class C airspace is relatively easy because it is much smaller than Class B airspace. Watch out nevertheless. These areas surround airline airports and still contain quite a bit of air traffic. Many, if not most, contain secondary airports that lie underneath the floor of the outer layer, and a good number of these are controlled airports. If you don't want to use the services of ATC for whatever reason, it's best to fly totally around the Class C airspace rather than underneath the outer tier.

Flying over Class C airspace is more practical than Class B because of the 4,000-foot AGL ceiling. Most general aviation airplanes perform well enough to overfly Class C airspace by a good margin without pressing the need for pilot oxygen or getting so high that the performance of the airplane starts to suffer. This is especially true east of the Rocky Mountains.

Staying well above the Class C airspace and below 10,000 feet will put your airplane in the altitudes where the 250-knot speed limit is still in effect, so you won't be confronted with civilian jet traffic operating at tremendously fast speeds.

Another twist to operations within Class C airspace is the requirement to communicate with ATC even if you're departing a secondary airport that lies within the lateral boundaries of Class C. The two-way radio contact is to be established as soon as possible after departure. The Class C airspace at many locations is so small that you'll be close enough to the radar site located on the primary airport to be able to talk to ATC while still on the ground at another airport located underneath the second tier.

If you can hear the controller, he should be able to hear you. Give a listen to the appropriate frequency (displayed on the chart) before you take off. If you can't communicate before you lift off, you won't need much altitude to be in the line-of-sight communication range, so be ready to start talking as soon as possible.

FAR 91 requires the same aircraft equipment be installed and used for operations within Class C airspace as for Class B. That means that you need a two-way radio with the appropriate frequency capability, and a Mode C transponder. There was much gnashing of teeth among some in the general aviation industry when the Mode C rules were first announced, but the use of this equipment has helped considerably.

When you make your first call to the controller, you'll be asked to confirm whatever altitude the radar screen is displaying. The controller needs to know that the encoder is working properly in order to rely on it for accurate information. When everything is confirmed, you'll seldom hear the old request from ATC in the nonencoding days: "Say altitude."

No regulatory restrictions are placed on student pilots flying in areas of Class C airspace. Because of the busy nature of Class Cs, students certainly ought to receive adequate instruction and practice with their instructors on board before wandering off into Class C airspace. The same goes for certificated pilots who either feel totally uncomfortable or are out of practice in dealing with ATC instructions. The ATC system is weakened when a controller has to cope with an unprepared or unpracticed pilot.

The weather minimums for VFR flight within Class C airspace are the ones that pilots have been used to for years. You need 3 statute miles of visibility, and you are required to maintain certain distances from clouds. The distances vary depending on whether you're above, below, or beside a cloud. The dictates are to remain 500 feet below all clouds, 1,000 feet above, and 2,000 horizontally.

These are obviously minimum cloud separation rules, and the knowledgeable pilot gives more leeway than that. Remember that the reason for the 2,000-foot horizontal requirement is because there just might be an airplane flying along IFR through clouds at cruise airspeed. You'll have a lot less time to see and avoid an airplane cruising straight and level toward you than you probably will have if one is climbing out of the clouds.

If you're cruising at 120 knots and the airplane heading toward you is cruising at 120 knots, your combined closing speed is 240 knots, 6 nautical miles per minute, or a mile every 10 seconds. That 2,000-foot distance is roughly one-third of a nautical mile, so that gives just 3 seconds before impact in which to become aware of the presence of another airplane, determine that you're on a collision course, decide what evasive action to take, and execute the evasive action to provide clearance. Three seconds is not very long.

Special VFR minimums require only 1 mile of visibility, and the ability of the pilot to remain just clear of clouds. Special VFR is not authorized in many Class C airspace areas for reasons that should have become clear in the preceding paragraph. To determine if special VFR is allowed in a particular Class C airspace parcel, look at the sectional chart. If special VFR is banned in that area, there will be a blue box around the airport name with the letters "NO SVFR," which indicate that fixed-wing special VFR is not authorized.

Special VFR should only be used when departing a controlled airport, whether in Class C or D airspace, to escape a very localized weather condition, such as an isolated rain shower or snow shower. It should never be used for cross-country VFR flying because that is simply pushing the weather too far.

It's also usually easier to fly through a Class C area if you don't want to fly around it. When you have established two-way communications with ATC (which you must do before entering the Class C airspace), tell the controller that you want to fly through the Class C at whatever particular altitude and heading suit your needs. If the controller can't accommodate that precise request, chances are very good that she'll be able to handle you with only a slight modification to the altitude or route that you requested.

## Class D airspace

Class D airspace is made up of what was once called control zones and airport traffic areas. It is set up around airports with operating control towers that are not within either Class B or C airspace. These are commonly the busier general aviation airports that serve little if any airline traffic. We used to just refer to them as controlled airports. While they are controlled, the proper nomenclature now is to say that they are Class D airspace airports.

The vertical and horizontal boundaries of Class D airspace haven't generally changed much from when these areas were called control zones. They generally extend 5 statute miles in all directions from the airport and have a vertical ceiling of 2,500 feet AGL instead of the control zones' 2,999-foot ceiling. There might be lateral extensions of the 5-mile boundary to accommodate the room needed by IFR aircraft to maneuver to make instrument approaches.

Before entering Class D airspace, the pilot is required to establish two-way radio contact with the tower, and maintain it while within the Class D area. If a radio failure occurs after making contact under VFR, you may continue toward the airport and land provided that you can maintain basic VFR and you get a clearance to land from the tower. If the radio fails, how do you get a clearance to land? Get out the books and dust off the arcane system of light-gun signals.

We like to pretend that modern aircraft engines and radios don't fail, but they still do. A steady green light from the tower when you are in the pattern indicates a clearance to land. When you are on the ground, various signals are used for taxiing instructions; flashing green means that you may continue to taxi, and a steady red light means to stop. (If you don't know where you're going at the airport, there is obviously no way to get a progressive taxi from the ground controller.)

For a complete explanation of ATC light-gun signals, see Table 3-1. It's a good idea to have this chart of light gun signals available before flying into an airport that is controlled. Memorize the signals, and keep a precautionary copy of the chart some place in the cockpit where you can get it out to review in a hurry if the need arises.

*Table 3-1. Light gun signals as described in the Airman's Information Manual.*

| Color and type of signal | Movement of vehicles equipment and personnel | Aircraft on the ground | Aircraft in flight |
|---|---|---|---|
| Steady green | Cleared to cross, proceed or go | Cleared for takeoff | Cleared to land |
| Flashing green | Not applicable | Cleared for taxi | Return for landing (to be followed by steady green at the proper time) |
| Steady red | STOP | STOP | Give way to other aircraft and continue circling |
| Flashing red | Clear the taxiway/runway | Taxi clear of the runway in use | Airport unsafe, do not land |
| Flashing white | Return to starting point on airport | Return to starting point on airport | Not applicable |
| Alternating red and green | Exercise extreme caution | Exercise extreme caution | Exercise extreme caution |

The light gun is quite bright and can be easily seen while in a normal traffic pattern or on the ground at an airport. But be aware that the beam of light that the signal gun emits is rather narrow; therefore, the controller has to aim the gun directly at you for you to see the signal. The signal device is called a gun because it looks like a big pistol and even has a gun sight type device on it to help properly aim the beam of light.

Departure requires the same contact with ground control and tower that we talked about at length in chapter 2. This radio contact must be maintained until clear of either the lateral or vertical boundaries of the Class D airspace.

Flying either over or around Class D airspace is not difficult due to the reduced size of the controlled area. It's generally safe to plan on overflying Class Ds because they only go up to 2,500 feet above the airport. Most cross-country flight will be conducted well above that altitude anyway unless lower cloud ceilings dictate a lower cruising altitude.

Don't press the issue and overfly at 2,550. Use good judgment, and try not to fly over at less than 2,000 feet above the ceiling of Class D (not less than 4,500 feet above the ground). If you can't cruise that high for some reason, just deviate your heading some and circumvent the Class D airspace altogether, or give the tower a call and announce your presence and intentions.

Student pilots are permitted in Class D airspace. Recreational pilots are not allowed, again because recreational pilots can't operate in any airspace where contact with ATC is required. A good deal of flight training is conducted from airports within Class D airspace. Because a student might be learning to fly from an uncontrolled airport, the certification sections of the regulations require at least three solo takeoffs and landings to a full stop at a controlled airport before becoming eligible to be licensed as a private pilot.

The VFR weather minimums for flights in Class D airspace are the same as for Class C. You must have 3 statute miles of visibility, and remain 500 feet below, 1,000 feet above, and 2,000 feet horizontally from all clouds. Most Class D airports do authorize the use of special VFR, but the same caveats apply as stated earlier. Special VFR should only be used in very limited circumstances and never as a substitute for general VFR minimums when protracted flight will be required in the poor weather conditions necessitating the special VFR clearance.

There is no requirement for either the installation or use of a transponder within Class D airspace *per se*, but if you have a transponder, it must be turned on. There aren't many airplanes left these days equipped with an electrical system that don't also have transponders. If yours doesn't, you're limited to flying only in Class D, E, and G airspace, and that eliminates a whole bunch of places from where you can fly.

Older classic airplanes without engine-driven electrical systems can operate in Class D airspace, but the classic airplane pilots still are required to have a two-way radio. With the advent of hand-held radios and with an external antenna mounted on these older airplanes, they can be and are regularly flown into airports in Class D airspace.

## Class E airspace

Class E airspace is described as general controlled airspace. What this means is that it is the airspace within which controlled flights may operate, and those are made by aircraft operating under IFR. Because the airspace in Class E is called controlled does not mean that all of the traffic flying in it is actually controlled. Class E airspace is where the greatest portion of VFR cross-country flight occurs, guided solely by the pilot's desires, complying with the operating rules contained in FAR Part 91.

Class E airspace begins at the ceilings of either Classes B, C, D, or G. Within a 5-statute-mile radius of an uncontrolled airport, Class E will start at the surface. All Class E airspace extends from its floor up to 17,999 feet. Remember that at 18,000 feet, you are entering the positively controlled arena of Class A airspace.

When flying through Class E airspace, there is no requirement to talk to any controllers, unless you are in a restricted area (we'll have more to say about that in a few moments). Class E is where you're free to go as you wish, as long as you follow the basic Part 91 regs.

Class E airspace has no special aircraft equipment requirements for flight at altitudes below 10,000 feet MSL. No radios or transponders are needed as long as you stay within Class E and don't get into one of the more restrictive classes of airspace; however, if your airplane has an operable transponder on board, you've got to turn it on and leave it on, including the Mode-C altitude function, which encodes your altitude and displays it on the radar controller's screen. This requirement was enacted so that controllers in air route traffic control centers (ARTCCs) who are normally working en route IFR traffic can give more meaningful traffic advisories to the airplanes that they are controlling.

Before this rule went into effect, you would often hear the center controller advise an IFR airplane of a potential conflict with another aircraft and end the warning with the words "altitude unknown." The pilot of the IFR airplane had no idea whether the traffic being called was a helicopter down at 500 feet AGL or an unexpected jet up in the flight levels. Now the controller is generally able to give you at least a hint by saying "Mode C indicates 4,500 feet." They say "indicates" because unless the controller is talking to the other aircraft, the encoder's reading is not confirmed.

It seems as though encoders, particularly those known as blind encoders, are more apt to fail and give false indications than most of the other avionics equipment. So still keep a lookout in all directions and altitudes when a controller has to hedge the advisory by telling you that the conflicting aircraft's altitude is merely "indicated."

Above 10,000 feet MSL, you've got to have an operable altitude encoding transponder in Class E airspace and use it unless you are within 2,500 feet of the surface. This 2,500-foot window of no-transponder-required airspace exists so that airplanes without transponders can fly in mountainous areas within 2,500 feet of the surface, yet still above 10,000 feet MSL.

Student and recreational pilots are permitted to fly in Class E airspace without any special authorization or logbook endorsements other than the ones otherwise required by FAR Part 61 for the operation in which they are engaged. These other required endorsements deal mainly with cross-country flights when the student pilot must be properly prebriefed and have the flight planning reviewed by the instructor before a solo excursion.

Recreational pilots are not permitted to make flights beyond 50 miles from an airport where they have received instruction; therefore, they don't make cross-country flights. (The Experimental Aircraft Association petitioned for removal of the 50-mile limit for recreational pilots. That bar to cross-country flights by recreational pilots might be reduced or altogether eliminated; check the current regulations, ask an instructor, or contact the nearest flight standards district office for more information.)

(I would like to see the recreational pilot certificate be made more useful. Since that level of certification was enacted, only a figurative handful of the licenses have been issued. If we are going to revitalize general aviation, it's got to become less expensive and less complicated for a person who wants to fly simple airplanes in the uncomplicated environment of Class E airspace.

Recreational pilots ought to be trained and viewed as the private pilots of the 1950s. All that those earlier private pilots lacked over today's training was instrument experience, radio navigation, and controlled airspace skills. A recreational pilot could be very easily given the additional training to make simple cross-country flights, say up to a range of 150 to 200 miles, without increasing the burden of the training very much. Then that certificate could be used for more real recreational flying than just the local type operations now authorized.)

The basic VFR weather minimums apply in most Class E airspace: 3 miles flight visibility, and remain 500 feet below, 1,000 feet above, and 2,000 feet horizontally from clouds. When flying in Class E above 10,000 MSL, the cloud clearance requirements increase to 1,000 feet below and above clouds, and 1 statute mile horizontally from them. Above 10,000 MSL, the required flight visibility goes up to 5 statute miles. This increase in both the visibility and cloud clearance rules above 10,000 feet MSL exists because above 10,000 feet MSL the 250-knot speed limit is not applicable, and VFR traffic might just meet a jet flying faster than 250 knots.

If the workload of a controller permits, you will be able to get traffic alerts from either the ARTCC or other radar facility in the vicinity. Don't count on it from many ARTCCs because the workload handling their IFR traffic often precludes giving VFR advisories to transient VFR traffic that they are not actually controlling.

You'll have somewhat better luck getting advisories from radar approach and departure controllers, but keep in mind that their primary responsibility lies in separating the IFR and VFR traffic within their Class B or C airspace. They can serve you if they have the time and if it's a light traffic period of time, but they don't owe any duty to airplanes in Class E airspace like they do to the other airplanes that are actually within their jurisdiction.

## Class G airspace

Class G airspace is very limited in scope. It was formerly called uncontrolled airspace. While it is possible for IFR traffic to be in Glass G, it will be doing so only for takeoff from and approach to uncontrolled airports.

Class G generally lies beneath the floor of all Class E airspace, except the Class E within 5 miles of uncontrolled airports that have instrument approaches. In the situation of an uncontrolled airport that does have an instrument approach, the Class E airspace will extend down to the ground within 5 miles of the airport; otherwise, wherever there is Class E airspace, Class G lies beneath it.

The upper limit of Class G will be either 700 or 1,200 feet above the surface; the upper limit varies depending upon whether there is an IFR transition area present. It's easy to see that Class G airspace is very shallow in depth and is what some pilots call "scud-running territory."

Class G airspace is no place for cross-country flying because it's just too low to be safe for protracted flight in fixed-wing aircraft. For some reason still unknown to this rotary-wing pilot, helicopter pilots thrive down low. You'll generally see helicopters flying cross-country at about 500 to 1,000 feet above the surface, but they can do so more safely than can airplanes because of their ability to slow down when necessary to check out some upcoming obstacle. Helicopters can also proceed with caution, turn around "on a dime," or set down on any level spot if the weather deteriorates. Airplanes don't have that kind of flexibility and should not be that low unless the pilot is intimately familiar with the locale, terrain, and obstacles in the area.

Unless you're within a 10-nautical-mile radius of Billings, Montana's Logan International Airport, there is no requirement to have or use a transponder in Class G airspace. This is truly the province of the birds, and you're on your own. ATC can provide traffic advisories, workload permitting, the same as when operating VFR in Class E airspace. But because radar is also a line-of-sight device, you won't be picked up on the radar screens of many ATC facilities while in Class G. You're so low that the effective radar coverage at that altitude will be only a few miles out from the radar antenna, unless that antenna is mounted on a mountain peak or on the top of a tall control tower.

Weather minimums in Class G are virtually nonexistent. All that is required for airplanes is that you have at least 1 statute mile of flight visibility during the day for flights conducted below 10,000 feet MSL, and below 1,200 feet AGL, and remain clear of clouds. For those few parts of the country where Class G exists higher than 1,200 feet AGL, you need the typical 500 feet below, 1,000 feet above, and 2,000 feet horizontal cloud clearance between 1,200 AGL and 10,000 MSL. Above 10,000 feet MSL, the minimums are the same as for Class E: 1,000 feet below, 1,000 feet above, and 1 statute mile horizontally from clouds.

If you want to fly night cross-country in Class G airspace, the cloud clearance rules are the same, depending upon altitude, but the visibility goes up to a generous 3 statute miles below 10,000 MSL. Above 10,000 MSL, the needed flight visibility is 5 statute miles.

Recipe for disaster: night VFR cross-country at low altitude in weather that would force a pilot to stay in Class G airspace rather than climbing up into Class E. Chapter 11 has more about night cross-country, but remember that you're running a very high level of risk to be flying at night at about 500 feet AGL in 1-mile visibility trying to remain clear of clouds that will not be that easy to see due to the darkness. I question the sanity of any fixed-wing pilot who tries it.

That's about it for the classifications of United States airspace. The rules can look intimidating at first glance, but hopefully the foregoing explanations have helped a bit. We can't overemphasize the importance of knowing how to conduct VFR flights, and the operating procedures mandated in the various classes, specifically in B and C. Remember that most FAR violations are a product of ignorance; study the airspace system so you don't suffer the needless misery of an enforcement case against you. Even if the FAA doesn't catch wind of a violation, these rules do enhance safety, especially around large airports and busy airports.

## Special use airspace

We're not quite done yet. We've talked about the airspace classification system, but what we haven't delved into is what didn't really change, except for some modifications in chart depiction.

A lot of airspace in our country is designated as special-use, either as prohibited, restricted, warning, or alert areas. In addition, there are military operations areas and military training routes that can easily affect VFR cross-country flights. Let's take a look at them.

## Prohibited areas

Thankfully not much airspace in the United States is prohibited to entry by general aviation aircraft. But where you do see a prohibited area, it is just that—prohibited. Enter it and the FAA will come down on you seeking severe penalties in the form of a possible suspension or revocation of your pilot certificate, or maybe a hefty fine, which is called a *civil penalty*.

Prohibited areas generally surround sensitive governmental facilities, such as the White House, Capitol, Pentagon, and similar installations that affect national security. They are shown on sectional charts by means of a blue border, inside of which are blue vertical and horizontal lines. Stay away. There is no time during which you can fly into a prohibited area. ATC won't allow you in either, so don't bother asking for a deviation from the prohibition.

## Restricted areas

Restricted areas are sections of airspace where all civilian flight is not prohibited, but it is restricted during certain times and at certain altitudes. Restricted areas contain some sorts of hazards to flight, many of which are invisible to the average pilot. Such activities as gunnery ranges, missile test flights, flight test areas, or aerial gunnery usually cause the designation of a restricted area.

Restricted areas are shown on sectional charts by a blue border around the area, and the border has blue "tick" marks closely spaced together, projecting into the restricted area. The area will also have an identifying number shown inside of the border, such as "R-5503A." To find out more about a restricted area, look on the bottom of the sectional chart. Let's use R-5503A as an example.

The bottom of the Cincinnati Sectional contains a small table of special-use airspace depicted on the chart (Table 3-2). Figure 3-2 shows the restricted area itself. Notice that a restricted area (R-5503A) overlaps a military operations area (MOA). The MOA is not restricted, so what we first have to be concerned about is R-5503A. Now examine Table 3-2.

*Table 3-2. The special use airspace on the Cincinnati Sectional Chart.*

| Unless otherwise noted altitudes are MSL and in feet; time is local. Contact nearest FSS for information. †Other time by NOTAM contact FSS | | | The word "TO" an altitude means "To and including." "MON-FRI" indicates "Monday thru Friday" FL – Flight Level NO A/G – No air to ground communications | |
|---|---|---|---|---|
| **U.S. P–PROHIBITED, R–RESTRICTED, A–ALERT, W–WARNING, MOA–MILITARY OPERATIONS AREA** | | | | |
| **NUMBER** | **LOCATION** | **ALTITUDE** | **TIME OF USE** | **CONTROLLING AGENCY**\*\* |
| R-5503 A | WILMINGTON, OH | 4500 TO FL 600 | 0800-2200 MON-SAT† | ZID CNTR |
| R-6602 A | FORT PICKETT, VA | TO BUT NOT INCL 4000 | CONTINUOUS MAY 1-SEP 15 †24 HRS IN ADVANCE | ZDC CNTR |
| R-6602 B | FORT PICKETT, VA | 4000 TO BUT NOT INCL 11,000 | BY NOTAM 24 HRS IN ADVANCE | ZDC CNTR |
| R-6602 C | FORT PICKETT, VA | 11,000 TO BUT NOT INCL 18,000 | BY NOTAM 24 HRS IN ADVANCE | ZDC CNTR |

\*\*ZDC-Washington, ZID-Indianapolis

This table tells us that there are four restricted areas on the Cincinnati Sectional Chart, but the only one concerning us is the first one shown. It is located near Wilmington, Ohio, which is southeast of Dayton, the home of Wright Patterson Air Force Base. R-5503A is one of the restricted areas where flight restrictions are in effect during certain times and at certain altitudes.

The table tells us that the area is restricted between 4,500 feet MSL and FL600. So if we're able to stay below 4,500 feet, the inquiry is finished because we're not even going to enter the restricted area. If for some reason we have to fly higher than 4,500, keep looking at the table. The area is restricted between the hours of 0800 and 2200 (8 a.m. to 10 p.m.) local time Monday through Saturday. If our trip is on Sunday, the restricted area is of no matter, as it is also not effective any other day of the week if we're flying through the area before 8 a.m. or after 10 p.m.

If we need to transit R-5503A during restricted days and times, and at restricted altitudes, all is not yet lost. The table tells us that Indianapolis ARTCC is the controlling agency for this area, and we can look up the frequency and give them a call. If the area

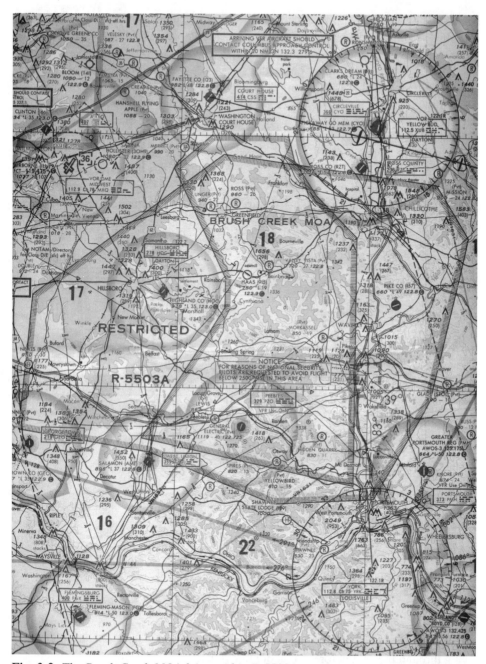

**Fig. 3-2.** *The Brush Creek MOA lying within R-5503A, as shown on the Cincinnati Sectional Chart.*

is not in use, we'll more than likely be allowed to fly right on through it. If the military is using it, for sure we'll be told that the area is "hot" or "active," and be denied entrance.

## Warning areas

Warning areas are airspace over international waters beyond the territorial limit of 3 miles. The activity in a warning area can be just as dangerous to flight as that conducted in restricted areas. The airspace can't be designated by the government as a restricted area because it is outside of the United States over international waters.

If you really want to know what goes on in a particular warning area, you can write to the FAA in Washington, D.C., and it will eventually write back and tell you. The best advice is stay out unless you want to play tag with a naval gun shell, fighter, or missile.

## Military operations areas

Military operations areas (MOAs) are established to provide separation between military training flights and *IFR* traffic. Depending on whether an MOA is hot or inactive, controllers will steer IFR aircraft away or allow them to fly through. As a VFR pilot, the decision is yours whether to penetrate an MOA.

Military flight training involves lots of high-speed maneuvers, aerobatics, intercepts, and other similar procedures that can compromise safety to other aircraft not participating in the exercise. While it's legal for a VFR pilot to proceed through an MOA, it's not necessarily too smart without knowing first if it's in use. The pilots of high-speed military aircraft who are playing Top Gun cannot keep a lookout for Cessna 172s plodding along through an MOA.

The bottom of the sectional contains a table of data about MOAs on the chart, much the same as for restricted areas. Table 3-3 replicates the information shown on the Cincinnati Sectional. Note that there is an MOA named Brush Creek, which is the one that overlies a good part of R-5503A that was previously discussed. Just the northern tip of Brush Creek MOA sticks out of R-5503A; otherwise, all of the MOA is within the restricted area.

*Table 3-3. The military operations areas (MOAs)
on the Cincinnati Sectional Chart.*

| MOA NAME | ALTITUDE OF USE* | TIME OF USE† | CONTROLLING AGENCY** |
|----------|------------------|--------------|----------------------|
| BRUSH CREEK | 100 AGL TO BUT NOT INCL 4000 | 0800-2200 MON-SAT | ZID CNTR |
| EVERS | 1000 AGL | SR-SS BY NOTAM | ZDC CNTR |
| FARMVILLE | 300 AGL TO 5000 | 0930-1430 & 1600-1700 MON-FRI | ZDC CNTR |
| PICKETT 1 | 500 AGL TO 6000 | SR-SS INTERMITTENT | ZDC CNTR |
| PICKETT 2 | 500 AGL TO 10,000 | SR-SS INTERMITTENT | ZDC CNTR |
| PICKETT 3 | 4000 TO 10,000 | SR-SS INTERMITTENT | ZDC CNTR |

*Altitudes indicate floor of MOA's. All MOA's extend to but do not include FL 180 unless otherwise indicated in tabulation or on chart.
†Other time by NOTAM contact FSS.
**ZDC-Washington, ZID-Indianapolis

Looking at Table 3-3, you'll see that the MOA is effective during the same days and times as is R-5503A, but the altitudes are different. A pilot can legally proceed underneath the restricted area flying below 4,500 feet during effective times on effective days; unfortunately he will still be in the MOA for about one-half of the geographical area covered by R-5503A. Indianapolis ARTCC is the controlling facility for the MOA; call it and see if the area is hot.

Some of the text in Fig. 3-2: "For reasons of national security, pilots are requested to avoid flight below 2,500 MSL in this area." That language appears in the bottom third of the area covered by the MOA. That is pretty strong advice to stay out.

When you fly over certain sections of the country, particularly in the South and West, there are wide areas of multiple MOAs around the bases where the military conducts much of its pilot training. Decades ago most military operations occurred at higher altitudes and did not much conflict with VFR general aviation traffic, but that isn't the case anymore.

A good deal of military flying is now done down on the deck, especially for Army helicopters and Navy, Marine, and Air Force attack fighters. Call the ATC facility that coordinates the activities in an MOA and find out what's going on before you plunge in. I cannot overstate the collision hazard that exists between military aircraft engaged in combat training and a complacent VFR general aviation pilot who blunders into their practice areas.

## Military training routes

The military also does training along what are known as military training routes (MTRs). These routes can be effective at virtually any altitude and time. They can be IFR training routes, which are called "IR Routes." They can also be flown VFR by military pilots, in which case they are called "VR Routes."

These routes are shown on sectional charts by the use of gray lines, along which will be depicted the route number. First, the number will be preceded by either IR or VR to designate whether it's an IFR or VFR route. The gray line is the center of the route, but each varies in width and can extend several miles on either side of the centerline. The numbering of the route gives us a good bit of information about the altitudes affected by its presence.

IR and VR routes that have no segment above 1,500 AGL will have a four-digit number. If any part of the route extends above 1,500 AGL, the route will have a three-digit number. This isn't a lot of help to the cross-country pilot because most cross-countries take place above 1,500 AGL. At least you can eliminate concern about the low-level MTRs with four-digit numbers.

When you see a three-digit numbered route, or a four-digit route if you're that low, contact a flight service station nearest the route to see if it's hot and at what altitudes. If it is hot, stay away. These routes are flown by all types of military aircraft, including bombers and fighters. While engaged in their training, military pilots have no absolute restrictions on their airspeed or other aspects of accomplishing their assigned missions, so don't assume that they will be flying slower than 250 knots. They are more likely to be operating at much faster speeds. Look out whenever you're near an MTR.

## Temporary flight restrictions

Every now and then the FAA issues a temporary flight restriction (TFR) to protect aircraft in flight and people on the ground because of some unforeseen event that poses a danger. TFRs can be issued because of a natural disaster, prison riot, or other happening on the ground that could be magnified or compounded by low-flying aircraft over it.

TFRs are issued by means of a NOTAM and are distributed through the FSS system. Usually a TFR is effective within 2,000 feet of the surface, but check each one that the FSS briefer tells you about to make sure that it doesn't have higher vertical limits.

TFRs are often issued as a result of some kind of disaster when either relief aircraft need exclusive access to the low-altitude airspace to get their work done, or something is occurring on the ground that could be made worse by low-flying aircraft. Due to the seriousness of the situation, the FAA gets very serious about enforcing the restriction and issuing violations against pilots who ignore a TFR.

Now that we've covered the various classes of airspace through which you'll fly during cross-countries, let's get on with the fun parts, and plan and fly some trips.

# 4
# Preflight planning and preparation

ALMOST EVERYTHING IN LIFE WORTH DOING IS WORTHY OF PREPARATION. The night that a major-league pitcher is going to start a game actually begins in midafternoon with a meal of high-energy foods to be properly fueled for the effort. Enough time will pass for digestion to get started and the pitcher will not feel stuffed. The exact timing is a matter of individual requirements, but each player has a pattern.

The surgeon who is going into an operating room has already studied the patient's chart and x-rays and knows what to expect when she takes the scalpel in her hand. A trial lawyer had better know his case, and his opponent's, before he ever walks into the courtroom. Would you want your health, life, or fortune to be in the hands of someone who is going to "wing it" without being as prepared as possible? The obvious answer is "no," but the aviation community seems to have a substantial number of pilots who take an ill-prepared and cavalier approach to cross-country flying.

Additional preparation is appropriate for longer flights. The preflight planning stage takes on even greater importance if the proposed flight will take the pilot over unfamiliar terrain, through an area of challenging weather, or to a destination where he has not been before.

On a longer flight, there is an increased chance that you'll be faced with something unexpected, which could range from the humorous to the dangerous. The best way to eliminate surprises is to prepare for and anticipate all aspects of the flight from the comfort of the planning table, whether it be in your home, office, or at the airport. The amount of planning should increase exponentially with the distance or complications of the trip. A good friend of mine completed a totally trouble-free trip with his entire family from Columbus, Ohio, to Alaska and back, covering some 8,700 nautical miles in the process. He took six months to carefully plan his flight and only about three weeks to fly it.

## IMPORTANCE OF PREFLIGHT PLANNING

FAR 91 makes the FAA's standpoint on the importance of preflight planning patently clear. It says in part, "Each pilot in command shall, before beginning a flight, become familiar with all available information concerning the flight. This information must include (a) For a flight under IFR or *a flight not in the vicinity of an airport,* weather reports and forecasts, fuel requirements, alternatives available if the planned flight cannot be completed, and any known traffic delays of which the pilot in command has been advised by ATC." (Italics added.) The next paragraph of the regulation concerns information about runway length requirements at airports of intended use.

There should be no doubt about the importance of the cited regulation. Failure to abide by its commands is often the heart of an FAA enforcement case against the pilot who stumbles into unexpected trouble on a cross-country flight, yet who could have discovered the potential problem and avoided it had she done the proper preflight planning. The Feds don't tolerate running out of fuel, getting dangerously lost, ignorantly stumbling into adverse weather that was plainly forecast or reported, or other calamities that just don't happen to the adequately prepared. Lawsuits that get filed after accidents in which someone is injured or killed often center and focus their allegations on the pilot's lack of preflight preparation.

There is no excuse for taking off into poor weather that is forecast or reported, but the pilot has failed to be informed about or, worse yet, ignores. Understanding weather forecasts is only one element of proper preparation. Some airplanes in the fleet are well within the permissible CG envelope at takeoff, and they can end up with the center of gravity outside allowable limits due to fuel consumption during a flight. A pilot who prepares the right way for a cross-country, or for any flight, knows if the airplane will be overloaded or out of its permissible center-of-gravity range during takeoff and throughout the flight.

Some experienced pilots flying alone to a familiar destination over a route that they have flown many times before in a high-performance airplane equipped with long-range fuel tanks might think that they are justified in taking off after a casual determination that the weather will be good and then getting the charts out when they are at cruising altitude. Seldom are they correct.

It is true that a high-performance airplane with its faster speeds will be less affected by adverse winds because the percentage of decrease or increase in ground speed caused by such a wind over the no-wind cruising speed will be far less than in a slow aircraft; full long-range tanks do increase the odds of making a trip without a fuel stop or with fewer stops. The capabilities inherent in higher performance airplanes do expand pilots' options and make it less likely that they will encounter certain surprises that will tax the ability of the airplane to make a given trip.

On the other hand, there is a right way and a wrong way to do things. Flying a fast airplane does nothing for a pilot to lessen the chances of encountering unforecast weather and might even increase that hazard; as more ground is covered every minute, it becomes obvious that the faster airplane will be moving through more different air masses in a certain amount of time than will a slower machine. When traveling faster, the rate of weather deterioration also speeds up, and the pilot has less time to analyze an unexpected situation.

If the weather does start going down, the pilot of a faster airplane has to get the situation under control all that much more quickly. The fast-plane pilot is potentially in a more dangerous predicament than a slow-plane pilot who would have more time to think, plan, and act out a solution for the same problem. Speed does nothing for a pilot who fails to check NOTAMs to see if any planned navigational facilities are out of service, or if the intended destination airport will be closed with an airshow in progress at the ETA. Like everything in aviation, all things are compromises.

An inadvertent encounter by a VFR pilot with IFR weather proves deadly far more times in a high-performance airplane than it will in a slower machine. Things go haywire, and the high-performance airplane gets out of control quickly, soon developing the graveyard spiral when the aircraft is a sleek retractable. If you ever make this mistake, a fixed-gear, slower airplane is much easier to control in this emergency. Fast airplanes can solve some problems and can exacerbate others.

Waiting to check weather and talk to a flight service station until airborne is amateurish unless absolutely necessary due to lack of a telephone at the departure airport or other location. If you've been camping in the wild, and leave from a remote airstrip, an airborne briefing is justified and should certainly be obtained. But it is silly to obtain the airborne briefing because you didn't want to take the time to adequately plan your flight before leaving home or your normal airport of operations. It crowds the FSS frequency with a lengthy set of communications that could have been done by phone. Very few pilots can pay close attention to a briefing while trying to fly the airplane as compared to a telephone briefing with flight service while seated at a table with the charts spread out and a notepad handy and pencil in hand.

If your flight will be carrying inexperienced passengers, you have additional duties and responsibilities toward them. FAR 91 requires that all persons onboard the aircraft be briefed on the fastening and unfastening of safety belts and shoulder harnesses. Don't assume that your passengers know this item. Tell them, and show them if necessary. Show them how to close and open the cabin doors.

Brief them about the hazards associated with leaving the airplane before the engine is shut down, and the propeller has stopped turning (engines and propellers for a multiengine airplane). If you've got a fire extinguisher in your airplane, point out its location and proper use to all passengers. If your windows can be opened in flight, make the appropriate cautions about opening them as placarded and stated in the pilot operating handbook. Think of all the things that a neophyte passenger might innocently do that could cause problems, and prevent them with a good briefing.

Most passengers are edgy about flying in light airplanes in the first place. While we who are pilots don't agree with the hazards that the general population perceives are present in general aviation flying, realize that fears are there, and be sensitive to them. A person's perceptions are his own realities, regardless of how you might feel about them.

Don't take a rookie passenger on a cross-country that will encounter known turbulence; people do get airsick much more readily in a small plane. Even if they don't get physically ill, very few people will be comfortable flying in turbulence in light-planes, so why ruin the experience for your nonflying friends.

Folks who are not accustomed to general aviation can become very anxious about IFR flying. I can remember when I was a corporate pilot back in the early 1970s, and one of the company's owners who hated to fly in clouds asked about every 5 minutes when we'd be in the clear. Even though he knew it was safe and had flown for several years in that airplane, he still didn't like flying in clouds.

I always did whatever was practical to alleviate his concerns and keep him in his zone of comfort. After a while, I was able to get him to take the controls on days when he and I were in the airplane alone. He was transformed from a businessman who first doubted the viability of general aviation as a business tool into a person who looked for an excuse to fly. Even though he never became a pilot, he's still a staunch supporter of aviation.

If it's necessary that the flight involve a long segment, ask your passengers to use the restroom before taking off. Remember that they might not think about it if their flying experiences are aboard passenger jetliners. The airlines have a restroom on board; you don't. Even without that problem, most folks aren't comfortable flying more than two to three hours at a stretch. Plan leg lengths to suit everyone's comforts, not just your desire to get where you're going as quickly as possible. Use decent airports, and avoid those that might make a passenger apprehensive, such as one ringed by mountains.

Explain to your passengers what you're doing. They don't know why you do a magneto check before takeoff; tell them that the engine has two complete and separate ignition systems, noting that the engine would keep running on a single ignition system if one went out. Point out the pretakeoff check verifies that both ignition systems are working properly. Explanations will help passengers understand safety factors that are built into an airplane.

In this day of home and business electronic technology, many people have become nuts for gadgets and sincerely appreciate knowing what's going on. If you sense that one of your passengers is such a person, invite him or her to sit up front with you. Then use the flight as a learning session, and let the passenger see how easy it is to navigate with modern avionics. Point out your checkpoints as they come into view; people become comfortable with any new endeavor much more quickly if they can be involved in it. Anyone who can master a personal computer can quickly grasp the elements of radio navigation.

If any passenger is apprehensive about becoming airsick, that person might be much more comfortable in the front seat. I have become uncomfortable in the back seats of lightplanes, and so have many other pilots I know. For some reason, people don't seem to be bothered as much by motion sickness in the front seat.

Naturally, pilot experience does affect the amount of time spent planning a flight, but not the thoroughness of the planning. There is a big difference. Experience enables a pilot to absorb the needed information more quickly and allows a pilot to do the planning steps more efficiently; experience doesn't lessen what is required. The pilot with less cross-country time should perhaps be more detailed in preflight planning by doing such things as marking the visual checkpoints to be used on the chart itself and making a written list of all radio frequencies in the order that they will come up throughout the flight. The veteran might omit these "bookkeeping" items if she chooses, as long as she knows the information before takeoff, and can quickly retrieve the data when the need arises.

## PREFLIGHTING STEPS

Chapter 5 will go through the entire process of planning a cross-country flight without using electronic aids to navigation. Regardless of the means of navigation to be used, all cross-country flights have many common planning steps. Learn to go through them regardless of whether your airplane is equipped to the hilt with black boxes or has none.

### Chart selection

Unless your cross-country is purely for fun and you only want to enjoy going somewhere without really caring where it is, the basic route of the flight will be predetermined by where you are and where you need to go. That much is obvious, but little else is.

Almost all VFR flying in the United States is done with the aid of sectional aeronautical charts. The 48 contiguous states are covered by 37 sectional charts, each of which is named by a prominent city that is within the chart's area of coverage. The scale of sectionals is 1:500,000, which lends a usable scale of 8 statute miles per inch.

This scale was more convenient in the days when we all used statute miles for figuring all calculations in navigation, except that wind speeds have always been quoted in knots. Now that most airplane airspeed indicators have the primary readings in knots, most pilots trained since the early 1970s tend to think universally in knots. If you're like I am, you still intuitively think in statute miles and then have to do the mental gymnastics to convert to knots.

Sectional charts are published by the National Oceanic and Atmospheric Administration (NOAA) about every six months or so. The edition number and effective dates of each chart are printed on the title panel of the chart as it is folded when you buy it. Make sure that you purchase a current chart. The importance of having current charts cannot be overstated. The FARs require you to have "pertinent charts" on board, and the FAA takes the view that out-of-date charts are not pertinent.

Old charts don't get the job done safely: new radio towers do get built; radio frequencies do get changed; new highways are always being constructed; new lakes and reservoirs appear; airports are constantly being closed and a few new ones built; special use airspace is added and removed; and other classifications of airspace are put into effect at different locations. In the event that an FBO offers you a chart that is out of date, buy a current one someplace else, and tell the operator that the old ones should be discarded.

Sectionals are designed for VFR navigation and therefore depict ground features that are useful for visual navigation. These charts show towns and cities in yellow; as long as your chart is current, they are pretty accurate in outlining the dimensions of populated areas. There is a legend printed on the back panel of the chart as it is folded when you first get it.

Study this legend in detail because it is nearly impossible to get to it when you're flying. You probably refolded the chart to show your route of flight, and finding the legend might be a herculean task in the cramped cockpit when you unfold the map, read the legend, and refold the map while trying to maintain course and altitude.

Don't depend upon the airport symbols to accurately depict what facilities are available at any given field. Consult the *Airport/Facility Directory* or *AOPA's Aviation USA* for a detailed description of what you can expect to find in the way of services at various airports.

Note that the obstruction symbols on the chart are of two types, one more prominent than the other. The smaller obstruction symbol, which is made of two lines converging into a point at the top with a dot at the bottom, is used for obstacles that rise less than 1,000 feet above the ground. For obstacles at 1,000 feet and higher, the symbol is a larger pictogram that looks more like a radio or TV antenna. The numerical values beside either symbol show the height of the top of the obstruction above mean sea level (MSL).

Because you should be flying along with your altimeter set for local station pressure (below 18,000 feet), a quick glance at the chart and your altimeter should resolve any question about clearing an obstacle. Immediately below the numbers that show the MSL elevation of the top of the obstacle, you'll see another set of numbers in parentheses; that number is the height of the structure above the ground (AGL). If you see the letters "UC" near an obstruction symbol, that means that it is under construction, and the elevations are not reliable; the obstacle should be given wide berth.

The legend also shows the different topographic data displayed on the chart. The symbols for highways, water features, power lines, landmarks, deserts, swamps, outdoor theaters, quarries, and other such identifiable things on the ground are of tremendous value for visual navigation.

At the top right corner of the legend, there is a depiction of how to read the airport data that is printed on a chart near an airport symbol. Blue airport symbols indicate controlled airports; magenta indicates uncontrolled fields. You should know how to read these airport data blocks in your sleep because they contain most of the vital information about any airport that you might need to use in an emergency, when you won't have time to check the airport directories for a full disclosure of what's there.

Another part of the legend shows all of the ways radio aids to navigation are displayed. Sectionals don't show certain radio facilities, such as instrument landing systems and other electronic aids that are pertinent only to IFR flight. Because sectionals are used in visual flying, this omission is of no great importance to the pilot who cannot fly IFR.

There is a good deal of information shown about each VOR station on the chart, and the station box also tells you whether there is an FSS located at the VOR, or whether communications with the FSS are by means of a *remote communications outlet* (RCO) from the VOR to an FSS located somewhere else.

The last portion of the legend deals with the symbology used to depict classes of airspace and special use airspace on the chart. Pay particular attention to the boundaries of Class B and C airspace, and the Mode C veils around Class B airspace because of the transponder and other equipment requirements in effect in these areas. Note that prohibited, restricted, warning, and alert areas are shown with blue boundaries around the affected areas, while MOAs are boxed with magenta-colored sides.

There are two other types of aeronautical charts published by NOAA for VFR flying, the terminal area chart, and the world aeronautical chart. Terminal charts are put out by NOAA for every terminal area that has a Class B airspace designation.

Terminal charts are extremely valuable if your trip will take you either into Class B airspace or anywhere near the 30-mile Mode C circle around the primary airport in any Class B airspace. Terminal charts have a scale of 1:250,000, so that the view is exploded at twice the size that the same ground area appears on the sectional. The legend is the same on terminal and sectional charts, except that the terminal chart shows more information.

A look at the Cleveland Terminal Area Chart (Fig. 4-1) shows additional information that is not displayed on a sectional. First, note that the terminal chart shows the approximate traffic paths in terms of departure and arrival routes for airline traffic in and adjacent to the Class B airspace. These are shown by means of blue lines along which there are pictograms of airliner-type airplanes. Departure routes show the airplane symbols headed out of the Class B airspace; arrivals are depicted by means of the airliner's symbol being headed inbound.

You will also see some numbers along these routing lines. The numbers indicate the usual altitudes flown by the airline traffic using those paths to get out of and into the primary airport. You should be clear of the airline traffic if you stay beneath the floors of the various tiers of the Class B airspace. But the wise pilot will be ever that much more vigilant when crossing the depicted arrival and departure routes used by the airliners.

Terminal charts also show visual checkpoints that are routinely used by VFR traffic flying into, out of, and around the Class B area. Find the small town of Vermilion,

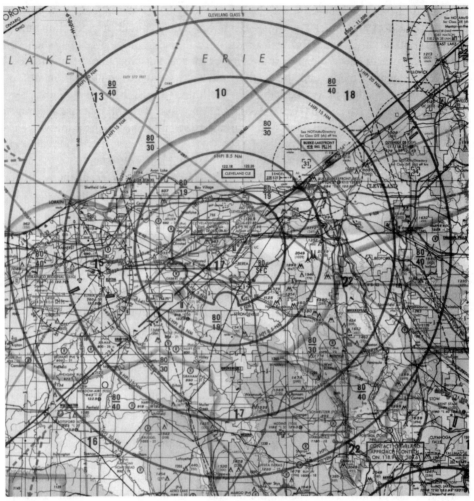

**Fig. 4-1.** *The Cleveland Terminal Area Chart is an example of the terminal area charts that are published for each Class B airspace area.*

Ohio, on Fig. 4-1. It's right on the shore of Lake Erie, west of Cleveland, just outside of the last ring of the Class B airspace. You'll see two things at Vermilion.

First is the little magenta-colored pennant-shaped flag. This is the symbol for an established and recognizable visual checkpoint to be used, as appropriate, when communicating with the controllers at Cleveland Approach Control. The other thing that you'll notice is the information box just west of Vermilion that tells the frequency to use to call Cleveland approach in this area. It's 124.0 MHz for us civilian pilots, and 360.6 MHz for military. If you look on the east side of the chart outside of the last ring of the Class B airspace, you'll see that in this area the civilian frequency is 125.35 MHz.

If you are circumnavigating Class B airspace, it's a lot simpler and safer to use a terminal chart because the scale of a terminal chart spreads everything out, making it easier to spot and use landmarks to stay out of the Class B airspace. If you don't have a Mode-C transponder, you can see that the 30-nautical-mile Mode C veil is also shown, and you should be better able to avoid it.

The third type of VFR chart is the world aeronautical chart (WAC). The use of WACs is falling off dramatically; a proposal was made to eliminate them altogether. The WAC is printed with the scale at 1:1,000,000, so each WAC covers twice the area of a sectional. Naturally, information and detail suffer.

WACs were extensively used a few decades ago by the pilots of high-performance piston-powered aircraft before many jets were in the air. Also in this era-gone-by, there was a lot more VFR flying by faster and higher-flying airplanes than today. Virtually all cross-country trips flown by turbine-powered airplanes is done under IFR these days, and the IFR system has its own sets of charts that serve the unique needs of instrument flight.

While they're still available, WACs serve at least one very useful purpose, and that is the initial planning of truly long flights. Because of their scale and the fact that it takes only 12 WACs to cover the 48 contiguous states, they're more easily managed than sectionals for plotting long-range flights. If you're planning a trip of around 600 miles or farther, it's very easy to use a WAC to identify the VORs, airports, and restricted areas along your route.

If you do your early planning using WACs, you'll need to transfer the information and course plotting to sectionals before actually flying the trip. The idea is that it's just easier to take your first looks at such a flight without having sectionals spread all over your living room floor.

It would not be wise to use a WAC for actual en route navigation. Because of the inherent limitations of their scale, less information is shown on WACs, and the symbols are too small for them to be suitable for VFR navigation by pilots of typical general aviation airplanes. If you fly a high-performance airplane and are going a fairly long distance, you might use a WAC for the en route phase of the flight, especially if you're using some sort of electronic navigation; you probably would be using it in a fast airplane.

But be sure that you have the pertinent sectionals at hand ready to use when you descend from cruise altitude and get near an airport where you want to land. You should have your sectionals already folded to show your planned route of flight along the entire trip. If you have any kind of emergency or other occurrence that calls for quickly knowing what's underneath you or calls for you to replot a course, forget the WAC and get out the sectional chart.

Another caution in the use of WACs also relates to their scale. Because the chart covers twice the area of a sectional, be on guard for the fact that things look like they're twice as close to each other on the WAC as they would appear on a sectional. One inch equals 16 statute miles on a WAC, and only 8 statute miles on the sectional.

Be extremely careful if you have always used sectional charts and take off on a trip with WACs because looks can be deceiving. Imagine that you start getting concerned about fuel or it's getting late in the day and you aren't current for night flying; you can quickly get lulled into thinking that an alternate airport is safely ahead if you don't realize that it's twice as far as it would be on a sectional's scale.

## Weather briefings

There is no other singularly more important factor in planning any flight than becoming familiar with the weather to be encountered. Even the professional airline pilot gets a detailed weather briefing before beginning every flight. Remember that airline pilots are not burdened with the need to remain under VFR at all times, as is the noninstrument-rated private pilot. Weather-related accidents fill volumes of NTSB reports, and weather accounts for the large majority of all general aviation accidents. To venture forth on a cross-country flight without an adequate knowledge of the weather ahead approaches playing Russian roulette.

The unfortunate part of weather planning is the fact that it can't really occur with any accuracy or dependability until relatively shortly before takeoff. Most pilots know that weather systems move more rapidly in winter than in summer, and as a general rule flying conditions in the United States are not nearly as reliable in the winter.

If you're making a cross-country from about mid-October to early April, be prepared to alter your plans at the last minute. Weather forecasting is not as accurate a science during that time as it can be during warmer weather when the systems move more slowly; the entire situation is more stable.

In the training curriculum for the private pilot certificate, you were exposed to the basics of meteorology. Like about everything else in aviation, a pilot should never stop learning about the weather and its effects on flight. The more you know, the more you'll be able to glean from the first step in weather briefings, which is to get the big picture.

There are several sources to become informed about the large-scale outlook of the weather across the country. Most newspapers have a weather map included with their forecasts. Look at the newspaper's national weather map, and note where the low- and high-pressure areas sit across the country. Locate the cold, warm, stationary, and occluded fronts. Most importantly, if you don't know the general characteristics of these different air masses and the frontal lines between them, it's time to hit the books again.

Most TV news shows have a weather segment that will show a national map. This display will probably be more current than what appeared in the morning newspaper because the newspaper map was prepared many hours before it hit your front doorstep. If you get it in your area, "The Weather Channel" on cable TV is an excellent source of the big picture because it usually shows not only the current surface weather map, but also forecasts about a day ahead to let you know where the systems are expected to move.

Because low clouds that might affect VFR weather minimums usually accompany precipitation, look for the areas where "The Weather Channel" meteorologists think that it will rain, snow, or sleet during the time that your flight will occur. If precip is shown, you need to get really detailed in the next steps of obtaining the needed weather data.

"A.M. Weather" is carried on many Public Broadcasting Service stations. The program is specifically geared toward the pilot and is possibly the best source of information of a general nature. As should be apparent from its title, this show is usually on the air early in the morning; it should be considered required viewing for any pilot who is planning to fly that day.

After you get the big picture, you might decide to cancel the trip right then and there. Don't be overly optimistic with yourself. If you're strictly a VFR pilot and a big low or a strong front is along or forecast to be along your proposed route of flight, it's probably best to call it a day and do something else. Hoping against hope that the bad stuff will somehow disappear has brought many a pilot to grief.

I cannot count the times that I have canceled short cross-countries that had to be flown VFR and then driven the trip only to see that I probably could have flown. But I'm always happier to go through that scenario than to be scud-running along on a journey that would have been much more comfortable down on the freeway in a car, or up in the flight levels in an airliner. I've left airplanes from Sault Ste. Marie to Tampa and come home on the airlines when I had to get back, only to go retrieve the little airplane later, but I got back.

## Preflight weather briefings

Three kinds of preflight weather briefings are available from the flight service station network: standard, abbreviated, and outlook. The standard briefing is the most detailed of the three and should always be obtained before each cross-country flight. Theoretically, standard briefings can be obtained up to six hours before takeoff. Unless the weather is forecast to be severe clear, you should not rely on a standard briefing that old without updating it shortly before the flight is commenced.

When you call the FSS to request any briefing, be ready to tell the briefer your aircraft number or your name if you don't yet know the call sign of the airplane you'll be flying. Also provide your route of flight and whether the flight will be restricted to VFR. If you don't give this information, the briefer will ask anyway. All of the calls to an FSS are recorded for everyone's protection; in the event of an accident or some regulatory violation, there won't be any argument about what was said by whom during the briefing process.

When you've gone through the preliminaries with the briefer, the standard briefing will automatically include the following steps and information:

*1. Adverse conditions.* The first step is for the briefer to alert you to any significant meteorological and aeronautical conditions that might influence you to alter your proposed flight plan, or maybe even to forget about it that day. These adverse conditions will include warnings about hazardous weather, runway or airport closures, and navaid outages.

*2. VFR flight not recommended.* When you have told the briefer that you plan to make your flight under VFR and IFR is not an option, the briefer will decide if you should be advised that "VFR flight is not recommended" for this route at this time. If

VFR flight is doubtful due to present or forecast sky conditions and visibilities, the briefer will first describe the conditions and the affected locations before advising that VFR is not recommended.

This caution is purely advisory in nature; the pilot in command is the final authority about whether to make any flight. But when you hear these words, you ought to listen up and discount the warning only after serious consideration. Some more experienced pilots think that the FSS folks advise against VFR too often. If you're a new pilot, or if you're flying in unfamiliar territory, give heed to such an advisory before you blast off VFR.

*3. Synopsis.* This stage of the standard briefing is a short statement describing the type, location, and predicted movement of weather systems or air masses that affect the route of flight. The level of your own weather knowledge and experience will play heavily upon your understanding of this stage of the briefing.

The well-seasoned pilot can get as much information from the synopsis as about any other portion of the entire briefing, while the analysis of weather systems might be Greek to a pilot who learned just enough weather facts for just as long as it took to pass the private pilot exam. As we said earlier, no pilot's education is ever complete, and if you ever want to be able to do a decent job of analyzing weather for yourself, the basic building blocks of weather wisdom come from knowing the elementary characteristics of weather systems and air masses and improving your knowledge from there.

*4. Current conditions.* Next you will be given the current weather at your points of departure and arrival, along with selected en route reporting stations. These observations are normally less than an hour old, but sometimes are an hour or two older. This is one of the most valuable steps in a good briefing. We all know that on occasion the weather forecast isn't very accurate.

When you get the next level of data from the briefer, the forecast en route, see how the weather that you've just been given in the current conditions matches up to what was forecast to be in existence at those places at that hour. There is no better check on the accuracy and dependability of a forecast than to compare it against what is actually happening at a given location at a stated time.

If they match up, then the forecaster had a good day. If they don't match up, watch out. If the forecast is for gradually deteriorating weather, but it appears that you can complete your planned flight before conditions get too poor, pay a great deal of attention to how the actual conditions compare to the forecast rate of change.

*5. En route forecast.* The briefer will tell you what the forecasting folks think the weather will be along your route of flight. This step should have three subsection forecasts: area, route, and terminal. The briefer should give you the area and route forecasts that affect your planned trip and should also throw in the terminal forecasts for selected airports along the way.

Area forecasts are general predictions of aviation weather within the boundaries of a large geographical area. Area forecasts come out three times each day and are amended as the forecaster deems necessary if things start to develop otherwise than as predicted. The continental United States is covered by six area forecast regions.

The route forecasts put out during the morning and at midday are valid for 12 hours; the forecast issued in the evening has an 18-hour period of validity. About 300 route forecasts are issued three times daily.

Terminal forecasts relate to a specific airport for which the weather is forecast. Terminal forecasts are also issued three times a day and are valid for 24 hours after the forecast comes out. About 500 airports in the 48 states have a terminal forecast.

*6. Destination forecast.* Next the briefer will tell you what is forecast for your destination at the approximate time that you'll arrive. Be sure to request the forecast for several hours before and after your ETA if the information is not volunteered. If anything is vital to weather analysis, it's the trend of weather condition development. You need to know if they are forecasting stable conditions for several hours, or if you're flying into an area of rapid change in the weather.

Because cloud ceilings are reported and forecast in terms of feet above the ground level of the airport, you need to know if the particular airport is atypical in any way from the surrounding territory. Is it in a valley or on a mountain top? The answer to this question certainly affects whether a forecast of a marginal ceiling should be cause for concern.

*7. Winds aloft.* You will next receive the forecast direction and velocity of the winds aloft in 3,000-foot increments. These forecasts also include the predicted temperature at each of the forecast levels. Temperature data is more important to the IFR pilot who might well be concerned with the possibility of icing in clouds or precipitation. But even the VFR pilot should note if the temperature is expected to be near or below freezing at any planned altitude during the flight if there is just the slightest chance that rain or other visible moisture might be encountered.

*8. NOTAMs (Notices to airmen).* NOTAMs are reports of such things as navaid outages, temporary hazards around airports like construction activity and runway closures, and perhaps an airshow in progress. Don't skip this step or let it be skipped. Once about 25 years ago I flew a trip to an airport where I often went, sometimes once a week.

To our surprise, the airport was closed for a few hours for an airshow precisely when we arrived. I've always enjoyed aerobatic flying in the air and watching it from the ground, but it wasn't much fun stumbling into an aerial demonstration while only a mile or two from the traffic pattern.

*9. ATC delays.* Delays that are occurring in the air traffic control system aren't usually of too much importance to a VFR pilot, but when they do happen, they're usually the result of the weather being down at major airline hubs like New York, Chicago, Atlanta, or Los Angeles. If the briefer warns you about ATC delays, inquire as to the reason. If it's weather-related, that ought to perk up your ears if you are headed to the area where the delays are present.

*10. Other information (upon request).* Ask the briefer to tell you about any other conditions that might affect your route of flight. These can be such things as activity along military training routes or in military operations areas, customs procedures if you're flying across an international border, and approximate density altitudes at selected airports.

There you have the steps in the standard FSS preflight briefing. The key to obtaining a good briefing is to ask all of the questions that you need answered to develop the big and little pictures that affect your decision of whether to fly, modify, or forget your planned cross-country. You have to receive *and understand* the information given out during the briefing.

Don't be bashful about asking questions, or requesting the briefer to go over anything a second time. These people who work in the flight service stations are vital links in the chain of aviation safety and do an excellent job of giving pilots good briefings. In the rare case that you get one on the phone who is in a hurry, is apparently having a bad day, or for any other reason seems to not be giving you the attention to detail that you think you need, be as inquisitive as you need to be to obtain a proper briefing with all the information that enables you to properly plan your flight.

In a few instances over the years I have terminated a briefing, waited a few minutes, then called back to the FSS. Fortunately, my second call was answered by a different briefer, and things then progressed normally. Before we leave the subject of preflight briefings, let's take a quick look at the other types offered by the FSS system.

**Abbreviated briefing.** If you have received a standard briefing within the past six hours, you may request an abbreviated briefing to update it. The purpose of an abbreviated briefing is to correct any information in the standard briefing with any new data that is more current than what you received a few hours earlier. In calling for an abbreviated briefing, be sure to tell the FSS specialist that you have already received a standard briefing and the time that you were briefed. Then you will be given whatever new information is available that will make your previous briefing current.

**Outlook briefing.** Outlook briefings are used for long-range planning when the planned time of departure is at least six hours or more in the future. Six hours might not seem all that far in the future, but in terms of changing weather, it can be an eternity. Every flight must be planned on weather information that is as fresh as can possibly be obtained.

An outlook briefing should be used for basic decisions of whether to make the flight at all. When there is an isolated area of bad weather that can be watched for the next few hours, perhaps that area can be circumnavigated if it continues to affect only a small area. Unless the weather is absolutely perfect for several hundred miles around you and you're planning a flight that is either local or a very short cross-country of, say, 50 miles or so, never depend on an outlook briefing as your sole call to the FSS for weather information.

**In-flight briefing.** In-flight briefings are not a separate type or format of briefings; the term is used to refer to any of the foregoing briefings conducted on the radio while in flight rather than by phone or personal visit to the FSS. As we said previously, in-flight briefings should only be requested when some good reason exists that makes it impossible to get your information over the phone or personally at the FSS.

These long-winded radio conversations tie up the communications frequency for an inordinate amount of time. In-flight briefings should be reserved for the situations when there is just no alternative. Remember that when circumstances dictate, take as much time as needed to get a thorough briefing over the radio as you would by any other means. A little inconvenience to others wanting to use the frequency is a small price compared to the possible consequences of continuing a cross-country flight without proper weather information.

## EN ROUTE FLIGHT ADVISORY SERVICE

En route flight advisory service (EFAS) is not a true step in preflight planning or briefing, but it is a valuable resource for cross-country pilots. EFAS is specifically designed as a service to provide en route aircraft with timely and pertinent weather advisories. EFAS is also a collection point for pilot-reported weather conditions, which can be among the best information available to any pilot.

Communications with EFAS are designed to be effective for all aircraft flying over the United States at altitudes above 5,000 AGL on one common frequency, 122.0 MHz. You need to look in the *Airport Facilities Directory* to determine the FSS that provides this service in the area where you are flying. The service is commonly called *flight watch*.

For instance, in the northern Ohio area you would initiate a call to Cleveland Flight Watch and identify your aircraft number and location. The FSS will respond, and you can then ask for whatever information you'd like. If you can't decide which facility to call, just use 122.0 MHz and make a generic call to "flight watch" without the FSS name, giving your aircraft type, call sign, and position. The appropriate FSS will answer.

EFAS is not designed for position reports, the opening or closing of flight plans, or for detailed in-flight briefings. Rather, it's a source of pilot-reported information of winds, weather conditions, turbulence, icing, and similar factual data provided by by pilots. The flight service specialists on the ground pass the information to other pilots in flight or on the ground during a preflight briefing.

Pilot reports are so valuable because they tell you what someone else has actually experienced rather than offering a forecast of conditions that might occur along your route. Before making a pilot report yourself, be sure to consult the *Airman's Information Manual* for the various definitions of terms that are used in a report to accurately describe conditions. Turbulence is often improperly reported.

The definition of extreme turbulence is that the airplane is out of control; a report of severe turbulence means that the aircraft is momentarily out of control, and is being subjected to abrupt changes in altitude and attitude, with large variations in indicated airspeed. If you encounter extreme or severe turbulence, it ought to be reported for sure.

Many pilots exaggerate their descriptions of turbulence out of ignorance of the real definitions; don't be among them. As long as the airplane is in positive control at all times, the worst level of reportable turbulence is only moderate.

Pilot reports are a vital part of the weather information system and should be made by all pilots. If you encounter anything different from what you expected based upon the forecast that you received during your preflight briefing, make a pilot report. Do so regardless of whether you're faced with better or worse conditions than you expected.

You can make a pilot report to any FSS. If things are really different from what was forecast, especially if dangerously so, don't hesitate to call an ATC facility and ask them to pass your upcoming pilot report into the system in the event that you can't contact an FSS.

One summertime pilot report is helpful to fellow pilots operating in the eastern United States. Most of the time during the warmer months some amount of haze will be restricting flight visibilities at the lower altitudes.

If you fly on such a day, and especially if you climb high enough to get out of the haze into cleaner air, report the altitude where you reached the top of the haze. I've always appreciated knowing where I can expect to get out of the muck and into better visibility. This pilot report can be a welcome part of any pilot's preflight briefing if you've taken the brief amount of time that is necessary to pass this information on to an FSS.

## CHECKPOINT SELECTION

Every VFR cross-country flight should be planned to use visually identifiable checkpoints on the ground to verify both position and progress throughout the trip. Several factors influence what points on the ground should be selected as checkpoints during the preflight planning process.

The first of these factors is pilot experience. Those of us with a few thousand hours of cross-country flying under our belts can safely plan flights using checkpoints spaced more widely apart than would be wise for a newly certificated private pilot to rely upon; however, there is an initial period in every pilot's career when certain things about flying are intimidating.

To some of us (author included), the fear of getting lost loomed large in every one of the first few cross-countries that we made. Others have no trouble at all with that, but are afraid of stalls, crosswind landings, or other facets of routine flying that quickly pose no problem with a little experience.

If you are relatively new at cross-country flying and are doing your flying in an airplane that cruises in the neighborhood of 110 to 125 knots or so, plan your checkpoints about 10 minutes apart where the terrain and man-made features on the ground allow. At a groundspeed of 120 knots, 10 minutes will equal 20 miles. That's about right for a neophyte cross-country flyer. As you gain more hours, and if the visibility is good, you might eliminate every other checkpoint so that your designated ones are maybe 40 miles apart, which translates to 20 minutes of flying time apart.

When the visibility is poor, I prefer checkpoints as close together as I can reasonably choose them. When you can only see 3 or 4 miles ahead, a checkpoint 20 miles in front of you is as useless as if it were in the next state. If you drift off course very much before you get there, the chances are great that you'll never see the longer-distance checkpoint.

If the weather is good, and especially if you're flying a faster airplane, don't crowd yourself with checkpoints too close together because that can be as confusing as too few. Like most things about aviation, common sense and good judgment developed with experience will soon sort out this process for you.

Many pilots choose inappropriate checkpoints. Find features that are easily distinguishable from others. In Minnesota, one lake might look like the other thousands in that state. In Ohio, where there are no natural lakes other than Lake Erie, the few man-made lakes are fairly large and make very good checkpoints. Freeway intersections are wonderful checkpoints in rural areas, but in the Los Angeles area or New York City area they are probably worthless because only a pilot living in that respective area can tell one from another.

Out in the hinterlands, small towns are great choices as checkpoints. Airports are fine as long as they are small airports and you can fly close enough to identify them without getting too low or too close to inadvertently penetrate some type of airspace where you shouldn't be.

It's tough to pick out checkpoints in sparsely populated areas. You might go 50 miles or more between towns, airports, freeways, or other prominent features. This scenario can exist in mountains, over deserts, or when flying in the vast wooded areas of the northern United States. When you're flying in these parts of the country, you had better be excellent at your dead reckoning skills, be flying an airplane with redundant electronic aids to navigation, or maybe be smart enough to consider this kind of flying to be in your future, not your present.

It might seem amateurish to a seasoned pilot, but remember that freeways are not only checkpoints, but are also great routes along which to fly if you cannot find plenty of checkpoints on a long trip or on a certain segment of a long trip. I'm not ashamed to admit that I've flown that way many times, and I would again if circumstances dictated it.

## ROUTING OPTIONS

The selection of the route to fly to accomplish a cross-country seems to be a nonquestion in most instances, and it probably is. Most flights occur between two airports along a relatively straight line between them. Before you make that assumption, study your chart. Look for artificial and real obstacles in the straight path.

First of all, see if there are areas of airspace that should be avoided. A Class B or C airspace area is no tremendous problem, but unless absolutely necessary, try to stay as far from the primary airport as is reasonable. The closer you tell the controller that you want to fly to the primary airport, the more trouble the controller is going to have accommodating your request.

Look for military operations areas and military training routes. If your investigation discloses that either is in use, stay away, particularly if they are used by the military for flight training or other low-level flying at the altitudes where you're likely to be. Remember that a fighter pilot screeching along at high speed and engaged in combat training probably won't ever know you're in the way until too late to avoid a collision. Unless you're extremely vigilant, the same goes for the general aviation pilot trying to see and avoid an F-16 that is probably painted in camouflage and flying at several hundred knots.

What kinds of terrain are you comfortable flying over? Consider a direct routing from Grand Rapids, Michigan, to Milwaukee. It involves overflying Lake Michigan for more than 75 miles. Going around the lake to the south will add at least an hour in a typical lightplane; to me, that decision is a no-brainer. I do not fly that far over cold open water in a single-engine airplane if there is any other halfway reasonable alternative. Some people do without thinking twice about it.

One of my friends left Oshkosh headed for New York and flew right out over Lake Michigan. The route might not have been too risky because the annual Oshkosh event occurs in the heat of summer when the water is the least dangerous in terms of causing quick hypothermia and when there are at least a few boats out. Regardless of those factors, I probably wouldn't do it. I don't like overwater flying in singles, especially if they have fixed landing gear.

A ditching in a fixed-gear airplane will almost certainly result in the airplane going inverted as the gear digs into the water, greatly increasing the chances of physical injury to the occupants during the sequence. Then even if everyone is uninjured, they have to get out of a sinking airplane that is upside down. Those are not good odds for survival. A retractable-gear airplane will probably stay upright during a ditching in calm water, but you'd better at least have life vests on prior to touchdown if you expect to survive until picked up by a boat. A lot more about overwater operations is in chapter 13.

The next kind of terrain to present its own special considerations is the desert. In the southwestern United States, and in the eastern portions of southern California, there are hundreds of miles of open desert confronting cross-country fliers choosing direct routes between major cities. The first thing to remember about desert operations is that the high temperatures produce high density altitudes, with all of the negative effects on airplane performance that come as a result.

Navigation by pilotage can be nearly impossible in desert areas, due to the lack of populated areas and man-made features on the ground. Checkpoints are few and far between. Unfortunately, due to their great expanse, deserts are not as avoidable as are the Great Lakes. Most pilots will choose to fly over the deserts during the early morning hours, when the temperatures are lower and the convective turbulence is not as bumpy, nor does it extend as high. When the sun heats the surface and the normal turbulence gets going, you probably won't be able to climb out of it in a normally aspirated lightplane because the rough air can extend higher than 10,000 feet, especially in the summer.

If your route takes you over the desert, use a little more thought in your survival planning. We don't talk enough about planning to survive forced landings. Most pilots don't give much credence to the possibility of engine failure because a properly maintained modern engine is very reliable. One of the good aspects of a forced landing in the desert is that the landing itself can most likely be accomplished with relative ease and safety because most desert areas are relatively flat, and the airplane will probably survive intact. Then you have to survive until found. There are a few ways that you can enhance those odds.

## Search and rescue

First, activate your aircraft's emergency locator transmitter (ELT) after landing. ELTs have saved more lives in the last 20 years than about anything else could have for those pilots and passengers unfortunately forced down in rugged or desolate country.

If the ELT's antenna is intact and upright, the transmitter's emergency signal will be picked up by either an American or Russian satellite and the coordinates of the "hit" will be automatically forwarded to the U.S. Air Force Rescue Coordination Center. The entire search-and-rescue system, including an airborne search by the Civil Air Patrol, will be activated.

If your airplane is equipped with either a loran or GPS receiver, and if you have time during the glide down, and if the airplane's master switch can safely be turned back on after the landing, use the latitude/longitude readings from either unit to find your present position, and jot it down.

If the master switch can be safely activated, turn on one communications radio only (to save battery power), and try a call in the blind on the international emergency frequency of 121.5 MHz. If you have time during the descent, it's a great idea to make such a call before the landing; if you're heard then by either a ground station or another airplane, so much the better. If you get a response, you can now easily relay where you are from the position that you got from your loran or GPS. Don't overlook the value of prominent landmarks that might be nearby despite the desert terrain, and don't overlook the value of VOR radial intersections as locators for rescue efforts.

Many pilots carry a hand-held communications radio. If you don't, get one before any lengthy trip over the desert. While it doesn't have the transmitting power of the radios in the panel, a hand-held can be a lifesaver. If you carry instrument navigation charts, look up the air route traffic control center frequency that is listed for the area where you are, and try a call in the blind. At the very least, you will likely raise another airplane flying in the vicinity that is working that frequency.

Next, you've got to stay out of the sun's direct rays. If your airplane can serve as a shelter, great. But again, a little planning for extended desert flying will include a small shelter, usually in the form of a light tent. Thought beforehand is the key to living in the desert for a while.

Lastly, you'll need water and survival food. Every flight over the desert ought to carry a water container that can withstand a little beating around. Don't carry your water in something like a glass-lined Thermos bottle that could easily break and leave you with the two alternatives of either not drinking or ingesting glass shavings. Conserve food and water, and you'll make it until you're picked up.

The main rule in survival is not to leave the airplane unless absolutely necessary. It can be spotted by rescue personnel much more easily than can a person wandering about. The airplane is shelter, and its fuel and oil can be the source of a signal fire if appropriate. If the airframe is relatively intact, and if you conserve the battery, you also have a means of communication if you stay with the aircraft.

If you venture out away from the airplane aimlessly, you'll probably get lost, and the chances increase exponentially that you'll never be found. If you know for sure that there is a highway, town, airport, or other populated area close, leave a note with the airplane describing where you went, and the date and time that you left the aircraft.

Another facet of cross-country flying that presents its own unique problems is flight over heavily forested and unpopulated areas. These factors face pilots who fly in the northern areas of the U.S. and in Canada. While this book is not focused upon survival tactics, a wise pilot who is planning a trip over such terrain would take a look at the Canadian regulations.

Because so much of our neighboring nation to the north is sparsely populated and wooded, their rules require survival gear, food, and firearms to be carried onboard cross-country flights in many of the western provinces of Canada. Doing likewise in similar areas south of the Canadian border makes good sense, too.

## USING A NAVIGATION LOG

When you were a student pilot, your instructor undoubtedly had you learn how to use a navigation log. See Fig. 4-2 for an example of a blank navigation log. Regardless of your level of experience or certification, a navigation log is one cross-country tool that should remain a habit as long as you fly.

For short cross-countries of 100 miles or less, the experienced pilot might omit the step of completing and then using a log of the course, checkpoints, distances, and times. But a pilot who is newly licensed or otherwise not an old-timer at cross-country flying needs to continue the use of a log regardless of the length of the trip. These logs can be purchased from commercial sources, usually at pilot shops, or you can easily develop your own, especially if you have a personal computer.

In this modern day when almost all cross-country flying is done with electronic aids to navigation, some pilots think that the steps and time spent completing a navigation log are wasted. They will eventually get lost, become unexpectedly low on fuel or daylight, or suffer some other problem during a flight that could have been eliminated had they taken the few minutes to properly plan their flights. Proper planning includes filling out and using a navigation log.

If you think about it, you'll see that we call avionics "aids to navigation" for good reason. The black boxes do make the job of getting from here to there much easier than it used to be, and no one wants to go back to the days when the only means of navigation were by pilotage and dead reckoning only. But radios and aircraft electrical systems do fail, and if you haven't prepared your flight to be completed by really navigating yourself, without the help of avionics, a radio failure will cause some anxious moments. If you're prepared, it's no big deal, and you'll feel great about yourself at the end of the trip.

| Checkpoints | Altitude | Mag. Hdg. | Fuel | Dist. | GS | Time Off |
|---|---|---|---|---|---|---|
| | | | | | | |
| | | | Fuel | Dist. | GS | Time Off |
| | | | Leg | Leg | Est. | |
| | | | Rem. | Rem. | Act. | ETE |
| | | | | | | ATE |
| | | | | | | |
| | | | | | | |
| | | | | | | |
| | | | | | | |
| | | | | | | |
| | | | | | | |
| | | | | | | |
| | | | | | | |
| | | | | | | |
| | | | | | | |
| | | | | | | |
| | | | | | | |
| | | | | | | |

NAVIGATION LOG

ROUTE OF FLIGHT:

**Fig. 4-2.** *Navigation log.*

One of my airline pilot friends once told me that in airline training the goal is to practice emergency procedures until they become so ingrained in a pilot that they cease to be emergencies and are nothing more than additional procedures. General aviation pilots are wise if they take the same approach to training and flying.

In our next chapter, we're going to plan and fly a cross-country without the use of any electronic aids to navigation. We'll go through this exercise for two reasons. One, classic airplanes are becoming increasingly popular as a means of affordable aircraft ownership, and most don't have electrical systems or radios other than hand-held units. Two, regardless of how well equipped an airplane is, every pilot needs the practice of flying cross-country every now and then without depending on the modern wonders of electronic navigation. We'll complete and use a navigation log as the heart of the planning process. So let's get on with it and plan and fly a cross-country.

# 5

# A trip without black boxes

AERIAL NAVIGATION IS A COMBINATION OF SCIENCE, ART, PRACTICE, AND eventually judgment. From the time that the airplane first started flying away from its home base shortly after the machine was invented by the Wright brothers, pilots had to be able to get from one place to another in some semblance of organization. Pilots had to develop ways to do it without ending up, at least regularly, where they didn't plan to be.

Aerial navigation shares many common facets with marine navigation, from which it was developed. Bodies of water have currents through which ships sail; airplanes move in currents of air called winds. Both ships and aircraft need to be able to get from a point of departure to a destination without stopping en route to figure out where they are. But the speeds of a ship offer the more leisurely pace associated with open water navigation as compared to the much quicker determinations that must be made by the pilot of an airplane.

The system of gridding the earth with lines of latitude and longitude was perfected by sailors long before man ever took to the air. Early in the history of long cross-country flying pilots were trained in celestial navigation, finding their way using the stars and the sun for guidance as sailors did for centuries before man took to the air.

Thankfully, the star-shooting days are behind us, or I would have stayed with sports cars rather than aviation. But most all of the other methods of navigating with a clock, compass, map, and determination of speed are as valuable today for aviators as they were for sailors during the great sailing years and the times of the explorers. Before we go on, let's get some terminology straight.

Pilots often confuse the terms *dead reckoning* and *pilotage* and think them to be one and the same. They are not. Dead reckoning is the art of navigating by determining the direction to head using some simple mathematics to account for the effects of wind and the characteristics of the Earth's magnetic field on the path that the airplane will fly over the ground.

Use of the word "dead" in dead reckoning is actually a misnomer. The older, more correct term is "deduced reckoning," which means that by mathematical reckoning, a pilot deduces certain facts and makes certain assumptions. Somewhere in antiquity, somebody misspelled the shortened "ded" as "dead," and it has stuck since. Maybe that's why the graybeards often said that dead reckoning means that if you don't reckon right, you're dead.

Pilotage is the art of determining one's position by seeing things on the ground and deciding which way to go based upon whether your previous heading took you to the right checkpoint. Unless you're flying over featureless terrain, such as expanses of desert, forests, mountains, or open water, almost nobody today uses dead reckoning alone. There is always some pilotage involved in navigating without electronic aids, and we'll proceed with that assumption, but if you do venture into uninhabited areas, and if you ever go to Alaska, you'll fly over hundreds of miles of that kind of terrain. Be ready to reckon.

From here on, when we mention dead reckoning, we're talking about a combination of dead reckoning and pilotage techniques to ensure that we end up where we want to be. Every pilot needs to both learn and remain adept at dead reckoning for several reasons. First, radios and electrical systems do fail. Sure, they're more reliable today with solid-state radios in place of the old tube type, and alternators seem to last longer than generators at the job of providing the power that the boxes need, although when I review my airplane maintenance expenses over the years, I'm not so sure.

When things do go dark, you've got to shed electrical load quickly, especially if you need to save battery power to get the landing gear or flaps down for landing. If you need to fly for a while to reach a suitable airport to land, load shedding is crucial. That means that you might be faced with turning off a perfectly functional panel of radios before you run the battery dead, if the alternator or generator has called it a day.

Next, as we just mentioned there are places in the United States where there aren't many, if any, VOR stations, cities, or towns close together, or freeways to guide you. The western United States is sparsely populated, as is most of Canada, and so is quite a bit of the eastern part of our country, particularly in Appalachia.

Lastly, it's fun. When the weather is good, navigating by dead reckoning can be a pleasure and a confidence builder. When the clouds hang low and the visibility goes down, this kind of navigation naturally gets tougher, but is not impossible for the practiced navigator.

I started my flying in an Aeronca Champ back in 1965. This airplane had no electrical system of any kind, not even a starter other than someone's arms. After only two or three months, I decided that if a college student was going to afford flying, he had better figure out a way to do it other than pay the astronomical sum of $8 per hour to rent the Champ. So, for $600 I bought a one-half interest in a 1946 Taylorcraft BC-12D. It was only one year older than I, and it took me all over the country.

The requirement for private pilots to demonstrate radio communication and navigation skills had just come into the testing protocol, as had the emergency instrument flying portion of the private pilot flight test. So I had to rent a brand-new Cherokee 140 to accomplish the flight test and a few hours of training before it, but I never flew that expensive machine any farther than a few miles from its home base. The old T-craft and I went everywhere we could because it only consumed about 4 gallons of 35-cents-a-gallon gas each hour.

Interest in classic aircraft has seen a renewal in the past few years, even at the prices that well-restored old airplanes fetch today. This is the basic flying that generations of pilots fell in love with, and for pure enjoyment it still beats any 150-knot or faster retractable airplane fully festooned with black boxes.

Even if you fly a modern, well-equipped airplane, you've cheated yourself out of lots of kicks if you've never flown someplace without using all of the avionics, except for keeping the transponder turned on, as the regs now require if you have one.

In chapter 9, I spend a little time recounting once when a buddy and I got so lost in the Appalachians on the way back to Ohio from Florida that we literally didn't know over what state we were flying. But we got unlost and laughed about it for years after. That was my first real cross-country of more than a hundred miles or so. It took two days to go from Columbus to Ft. Lauderdale and another two to get back home. As you'll see in chapter 9, we finished the trip on the Greyhound bus, but that's another story.

This first trip to Florida began with the ink still wet on my temporary private pilot certificate after passing the flight test a few days before. I still remember about 90 percent of that flight as though it were just a few weeks ago. It began by leaving Columbus headed southeast toward Charleston, West Virginia. From there I started out across the mountains headed for Beckley, West Virginia, the first fuel stop. Then on into the piedmont country of North Carolina, and the first day ended in Waycross, Georgia.

When you're cruising at 90 mph into the ever-present headwind, you don't go far in a day in late winter when the days are still short. At that speed, you also spend an eternity over inhospitable mountains and the swamps of southern Georgia and northern Florida.

The second day went fine, and before it was over, the T-craft had carried me to Fort Lauderdale's Executive Airport, which was uncontrolled then. When I landed, I had total time in my logbook of 107 hours. After a few days in the sun and a couple of local sightseeing rides, it was time to head back to the cold.

The return trip was great, staying the first night in Georgia, and heading north to Columbus early the following morning. But we weren't far from my first experience with being "unsure of my position," which is pilotese for being lost. Those mountains in Kentucky, North Carolina, West Virginia, or wherever we were all looked the same. Chapter 9 explains how that little problem was solved.

After we crossed the Ohio River, the weather started "going down" fast. We got to Ohio University Airport, then in Athens, Ohio (they've since built a new airport with the same name and paved over the old one into a shopping mall or something), and decided that the better part of valor dictated calling it quits, even though we were only 60 or 70 miles from home. We checked the bus station, took out the last few dollars that my buddy and I had, and bought two of those first-class bus tickets that today's teenagers have never seen.

When we finally got to Columbus at about 10 that night, my cohort realized that his lightweight Florida jacket didn't carry the day there. He shivered for the half-hour that it took his grandparents to pick us up at the bus depot. Three days later I went back to Athens and brought the T-craft home.

As a total aside, having nothing to do with cross-country flying, I recently got the bug for another T-craft after owning a Bonanza, two Cessnas, an Aztec, and a Comanche. And I have flown a few thousand hours since 1966 in quite a few other airplanes, gliders, and helicopters. I called the FAA to find out where old N43324 was these days, and lo and behold it was owned by a gentleman just 100 miles from Columbus.

I called him but sank in my chair as he told me that he had bought it as a wreck to use for parts for another Taylorcraft that he had restored. He said it ran out of gas in Kansas and was totalled in the forced landing as it went through a fence. The only usable parts were one somewhat damaged wing and some fuselage tubing. It was a long drive home that night from the office.

Now that the hangar story is over, let's get down to planning and flying a cross-country without black boxes. It won't be as ambitious as my first trip to Florida with fewer than 100 hours total time when I started out. For your first such cross-country, let's plan to fly from the Fayette County Airport, located right outside of a town named Washington Court House, Ohio, to Elkins-Randolph County Airport, which sits on the south side of Elkins, West Virginia.

In Fig. 5-1, you can see a reproduction of a portion of the Cincinnati sectional chart with our course line drawn on it. For ease, we'll refer to Fayette County by the identification letters "CSS" for the NDB that is located just north of the field; we'll call the destination airport "RQY," which are the identification letters of its NDB that is located on the airport. So our route of flight is direct from CSS to RQY. Let's assume that we'll make this flight in a Cessna 172.

The first step in planning any cross-country, whether you'll use radio aids to navigation or fly without them, is to procure a current chart and lay out the course line between the two airports. Draw your course line with either a heavy pencil or a medium-point pen.

Don't be bashful about drawing on the chart as long as your marks and lines don't cover up important information displayed on it. I remember when sectionals cost 25

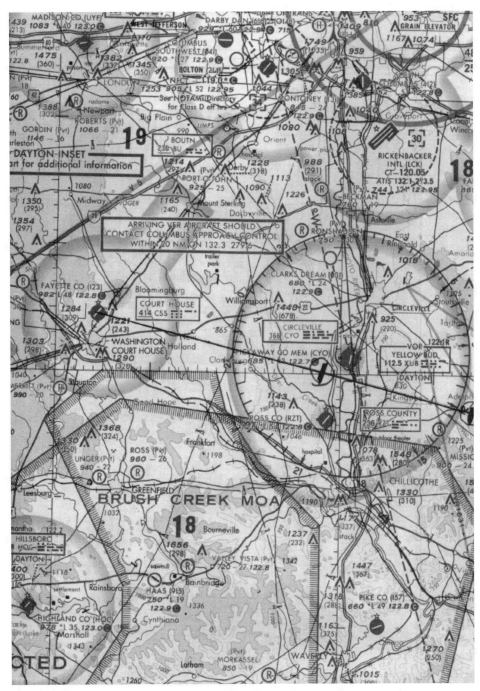

**Fig. 5-1.** *The course line from CSS to RQY with checkpoints circled.*

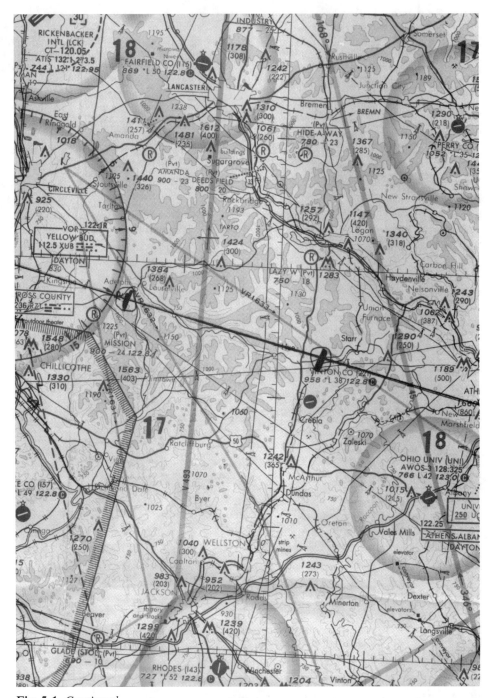

**Fig. 5-1.** *Continued.*

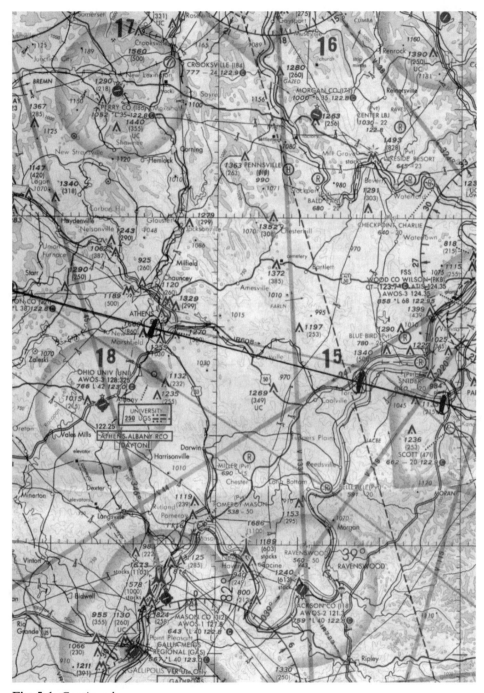

**Fig. 5-1.** *Continued.*

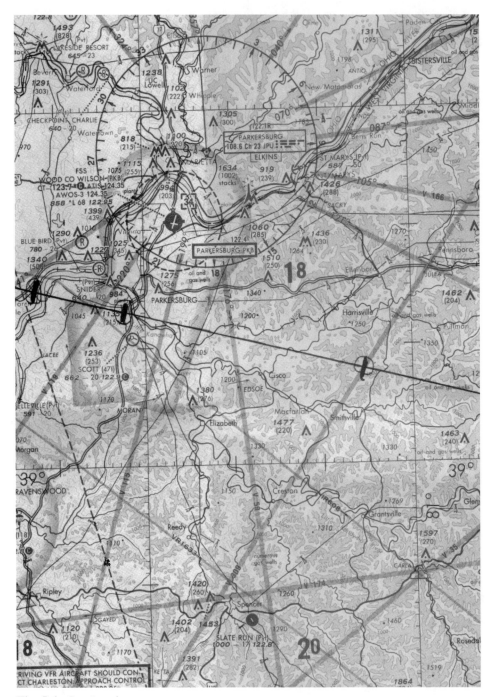

Fig. 5-1. *Continued.*

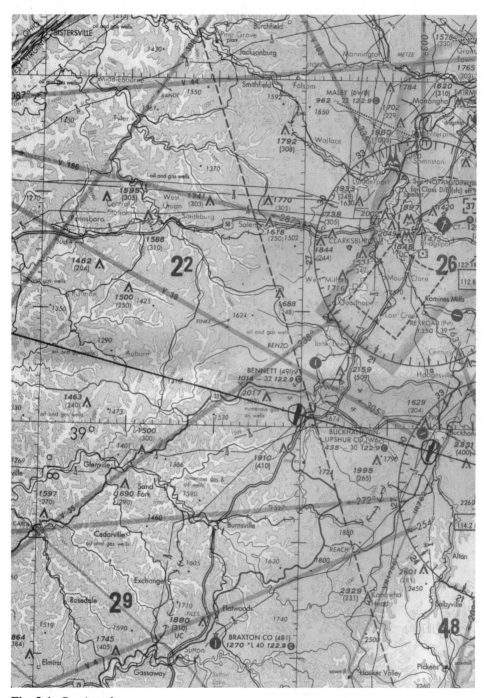

**Fig. 5-1.** *Continued.*

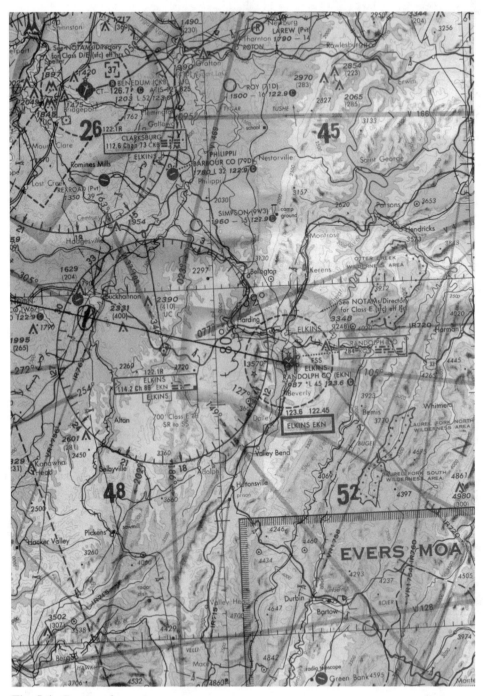

Fig. 5-1. *Continued.*

cents, and now they cost about five bucks. Disregard the cost; mark it up, and realize that the chart is replaced twice a year anyway.

When you draw your course line and have it on the chart so that you can see it (that's why you should draw it in heavy pencil or with a medium-point pen—so you can see it plainly while flying along), the next step is to select appropriate checkpoints to use throughout the flight to see if you've reckoned correctly and if you're still on course. When you pick checkpoints, two considerations will affect your choices.

First, space them appropriately. This means that you want them to be about 5 to 10 minutes apart as you fly over the route. Any closer together than that, and they become confusing and unusable; any farther, and you can get off course and not find the checkpoint if there's much wind drift on a day with poor visibility.

Second, learn to pick checkpoints that are easily identifiable. This takes some practice. Some things on the ground all look the same; other things are easily distinguishable. Look at Fig. 5-1, and you'll see that I've drawn a little tick mark inside circles at various intervals, usually about 15 to 20 miles apart. These are my checkpoints. In actual practice, you may omit the tick marks, but they're drawn here for clarity and ease of explanation.

Going eastbound from CSS toward RQY, the first checkpoint so marked is between two airports; one is just north of course at Circleville, and the other is a slight bit farther south of course called "Ross Co." Small, paved airports are excellent checkpoints. They're easy to see, and seldom confusing. So, for your flight, your first checkpoint is between these two airports. You want to be slightly closer to Circleville than to Ross Co.

It's about 20 miles from CSS to this first checkpoint, so you should have a good opportunity during this first leg to see how well your reckoning has produced a good heading to fly. If you're way off course, correct now because we're about to head into some less populated areas.

Our next checkpoint is about 15 miles farther away, and it lies between two small villages, Adelphi and Laurelville. Isolated small villages don't make good checkpoints very often because they're easy to miss and hard to identify as the correct place if you do spot them. But here you've got two close together, and you want to cross this checkpoint on the south side of Adelphi, and you should be able to see Laurelville just to the northeast of Adelphi. Note that you're in the neighborhood of VR 1632, a low-level military training route. Keep your head on a swivel for military aircraft doing high-speed, low-level flying.

Our next checkpoint is 19 miles away, 4 miles north of the Vinton County Airport. This isn't the greatest checkpoint because it's 4 miles from what you can visually identify. There isn't much else to use in that area, so we'll have to go with what there is.

Next, I've chosen the western edge of Athens, Ohio, as the upcoming checkpoint. It's a good checkpoint because Athens is a larger town, where Ohio University is situated. You want to be just over the south side of town as you fly by it. En route from our last checkpoint at Vinton County, you've passed over another low-level training route, VR 1633.

From Athens to the next checkpoint is 23 miles, until we arrive over the space between the Ohio River and a freeway that runs along the western bank of the river. Large rivers such as the Ohio are great checkpoints, so long as you can tell where you are along their courses. Here you have a good checkpoint coming into view because the freeway comes close to the shore with a 90° bend in the river from north to west.

After leaving this checkpoint, you'll be approaching the large city of Parkersburg, West Virginia. You want to pass just over the southern edge of Parkersburg and cross I-77, which runs north-south at the eastern side of the city. This makes a good checkpoint because Parkersburg is large and the interstate freeway is easily identifiable.

From here on, there isn't much to see except hills and trees for quite a while; therefore, if you've had any trouble keeping on course, you had better either have that problem solved, or land at the large Parkersburg airport and sort things out. The next identifiable object on the ground is 25 miles farther east, which is the town of Harrisville.

Our route will take us to the south of Harrisville, but you should see it 5 miles off to your left as you fly by it. If you don't see Harrisville, don't panic. In rugged terrain such as what you are now approaching, towns tend to be in valleys, and if you're flying low, the available slant-range visibility might not be enough to see a town tucked down out of view, unless you fly very close to it. That's why you want your heading nailed down before you leave Parkersburg.

After Harrisville, it's 33 miles until you'll cross I-79 near the south side of Weston. The leg of your trip from Parkersburg to Weston is about 58 miles long, and if you missed Harrisville, this segment could present an opportunity to get out of sorts with your planned position (lost). But if you've fine-tuned your heading by the time you get to Parkersburg, you should arrive near Weston just fine. Plus you'll cross a major interstate freeway at Weston; if you're a little north or south of course, you can adjust it there.

If you don't see Weston when you think you should, again, don't get upset. That freeway will be there, and if you've been looking, you'll see it as you cross it because it runs nearly perpendicular to your route of flight. If you've drifted far south of course between Parkersburg and Weston, you should see the large lake near Burnsville, some 12 or so miles south of where you belong. If your drift took you north of course, you'll be nearer to Clarksburg, which is big enough for about anyone to find.

In any event, get things sorted out here, and get back to the proper position near Weston if you've gotten off course. Also, if you did get off course by very much, it's time to figure out why. Many pilots can get off course because they are just not very good at flying a heading. Others do so because they didn't do the preflight planning of wind-correction angle and groundspeed correctly. Find out which error you made if things aren't going according to plan.

Now you'll head toward Buckhannon, just 15 miles away. I've put our tick mark between Buckhannon and its airport because only 2 miles separate the airport from town. This checkpoint should be readily apparent because the town is a larger village with an airport.

Now there are just 22 miles to go, but they can be challenging. Seven miles beyond Buckhannon puts you in the Appalachian Mountains for real. Notice that about 3 miles southeast of the Elkins VOR there is a symbol on the chart that denotes a point that's 2,720 feet above sea level. Five miles east of that is a ridge line, the first real one you've encountered on this trip. Right on top of that ridge, 4 miles west of RQY is the rotating beacon for the airport, and it sits at 3,570 feet above sea level.

This last leg can be tough if you aren't prepared and haven't studied the chart in advance because the terrain rises rapidly. The Buckhannon Airport's elevation is only 435 feet, almost 550 feet *lower* than CSS, from which you started this cross-country. The last 22 miles of this flight crosses a ridge of mountains that sticks up at least 3,570 feet, some 3,135 feet *higher* than Buckhannon.

After crossing that ridge, you have to descend to land at RQY, where the elevation is down to 1,987, or 1,583 feet lower than the ridge just 3 miles away. If you overshoot RQY, you're flying over mountains again because the terrain 5 miles east of RQY starts back up to nearly 4,000 feet.

Mountain airports are very often built in valleys, as is RQY. There's nothing particularly dangerous about operating from these airports if you keep your wits about you. In the eastern United States, as opposed to out West, the elevations aren't all that high, and the ridges can easily be crossed by almost any lightplane. But you've got to keep alert because more than one poor soul has flown into ridges while trying to take off from or land at these kinds of airports.

If the weather is at all questionable, *don't attempt the ridge crossing unless you're positive that you'll have enough ceiling above you to do it safely.* A trip that is cut short at Buckhannon surely beats trying to fly through the ridge line, instead of over it. Also, if there is much wind, be sure that you have enough altitude so that any downdraft on the lee side of the ridge won't pose a hazard.

## WEATHER PLANNING

Now that we've laid out our course from CSS to RQY, you need to see if today is the day to make the trip. What we've done so far is applicable to any day, whether it's today or next month. Now you've got to find out if this flight can be safely made today.

Before we call the weather, it's a good idea to set certain weather minimums that we'll need, and if they aren't there, we already have the good sense to call it off and try another day.

The first half of this cross-country isn't that demanding, neither in terms of navigation nor terrain. But if this is your first flight without the black boxes guiding your every turn, and if you haven't flown in mountains before, the second half from Parkersburg on to RQY is a different story. The necessary ceilings and visibility for the last half of this trip should be higher than for the first leg from CSS to Parkersburg. I would not want to make this flight without electronic aids to navigation unless the visibility were reported and forecast to be at least 8 miles. I would want the situation to be stable or improving.

As far as ceiling goes, remember that ceilings are reported and forecast with reference to height above the airport that is the subject of the report or forecast. So if the ceiling at Parkersburg is reported to be 2,500 feet, note that the Parkersburg Airport's elevation is 858 MSL. This means that the bottom of the cloud deck is about 3,350 feet MSL.

When you consider crossing that ridge line that sticks up to 3,570 feet just before arrival at RQY, the reasonably good ceiling at Parkersburg is about 200 feet *below* the top of the ridges west of RQY. This kind of awareness and planning makes the difference between a trip that is fun versus a trip that goes sour and results in a diversion to an alternate airport or worse.

The flight service station at RQY is an asset because you can get a good handle on the current and forecast conditions there. Because RQY's elevation is 1,987 feet, it's closer to the tops of the ridges than Parkersburg. I wouldn't consider crossing those ridges with fewer than 2,500 feet of space between me and the clouds if this flight were actually in a no-radio airplane; therefore, you'll want RQY's ceiling to be at least 4,100 feet to give you that clearance.

Pick out some good alternate airports because we all know that the weather prognosticators don't always get it right. This route is full of good alternates. Your first checkpoint was between two excellent airports. The third was abeam Vinton County Airport. Just south of Athens is my old friend, Ohio University Airport (although it's not in the same place as in 1966). Right after that, you could divert to Parkersburg and have great facilities available.

Parkersburg is a decision point for the reasons we've discussed above, but it is also the place to decide to call it quits if the weather is doubtful. After Parkersburg, the only real alternates are the two small airports at Weston (Bennett Airport) or at Buckhannon. If the visibility is going down, they could be hard to find. Also, you'll be going across about 60 miles of land that is without any good checkpoint other than at Harrisville, 5 miles north of your course. If you've drifted a little south of course, it could be obscured in poor visibility.

We need cloud levels at a minimum of 6,100 feet MSL and visibility at least 8 miles. If you want more visibility the first time you try such a flight, all the better, especially if the trip is for real in a no-radio airplane.

## WEATHER BRIEFING

Now that you know what weather you'll accept for this cross-country flight, you're ready to call the FSS and see if it exists. You should have minimum weather in mind for any flight, whether it's local or cross-country. There will be some cross-countries where you won't need ceilings as high as we need for the trip from CSS to RQY because you won't have the unique terrain clearance concerns that are present for this flight. Let's see what the weather is and what it's forecast to be.

You might not be aware that from CSS you should call the automated flight service station located nearby in Dayton, Ohio. You could go to the trouble and look it up in the AIM, but the FAA has solved this problem for all pilots. Wherever you are, all you have to do is dial 1-800-WX-BRIEF on any touch-tone phone to access the near-est FSS that serves that area; your call is automatically directed to the proper FSS.

If you don't have a touch-tone phone, you can still dial that number and be con-nected to the appropriate FSS, but you won't be able to use the automated features once you are connected; you'll have to wait for a live briefer. I prefer to talk to a per-son and seldom use the computerized methods of obtaining a briefing.

Once you've hooked up with Dayton FSS, you should know the types of briefings to request. You may ask for either a standard, abbreviated, or outlook briefing. You should request a standard briefing if you haven't received a previous one within the last six hours, or if you aren't making the return flight on a round robin, back into the weather through which you've recently flown. In those circumstances, an abbreviated briefing would suffice. The outlook briefing is for planning purposes only, when your planned departure time is at least six hours in the future. An outlook briefing should never be the sole briefing you get for any trip and should be followed by a standard briefing shortly before takeoff.

Tell the FSS specialist the type of aircraft you'll be flying, its registration number (or your name and pilot certificate number), the route of your flight, and whether the flight is limited to VFR or whether you can accept IFR conditions. Then the specialist will talk and you listen.

Have a notepad and pencil ready to take down what you're told, and jot your ques-tions down on the pad. Don't interrupt the briefing. Everything will probably proceed more smoothly if you wait until the end to ask questions or to request that the briefer repeat anything you didn't understand.

Now you've got to apply what you were told. Let's assume that the weather is go-ing to be severe clear with the lowest forecast or reported visibility at 10 miles. Plenty good enough, so you make the decision to go. Now you've got to go through some flight planning steps that you might not have used since your student pilot days. You've always let VOR radial tracking take care of such things as the wind-correction angle, magnetic variation, and deviation. Let's review them quickly.

Suppose that the briefer told you that the forecast wind aloft at 6,000 feet is from 320° at 25 knots, while at 9,000 it's from 340° at 35. You know that for the eastbound VFR flight you need to cruise at an odd altitude plus 500 feet. A flight at 7,500 feet makes sense because it complies with the rules and will give you at least the 2,500-foot clearance above the ridge just west of RQY. A flight at 5,500 wouldn't be high enough to meet that criterion. What wind will you encounter at 7,500?

You've got to interpolate from the data at 6,000 and 9,000. Because 7,500 is ex-actly one-half of the difference between those two reported altitudes, just take one-half of the difference in the forecast data. The winds can be expected to be from 330° at 30 knots at 7,500. Like any interpolation, this is only an educated guess, and you'll prob-ably have to fine-tune your heading en route using the checkpoints that were selected for this purpose.

There are many ways today to figure the wind-correction angle. You can take out your plotter, a pencil, and a sheet of paper and draw a wind vector like I learned to do

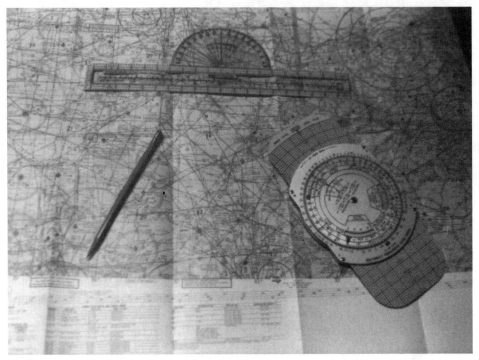

**Fig. 5-2.** *The tools used to plan a flight without "black boxes": E6-B computer, plotter, sectional chart, and pencil.*

in 1965. You can perform the same thing on the reverse side of the E-6B computer. Take a look at Fig. 5-2 to see the tools you'll use for this kind of flight planning.

But most pilots today use one of the many electronic calculators on the market that are geared to do aviation navigational problems. Whatever method you select is fine, as long as you do it. The wind correction technique that you choose will also calculate the groundspeed, which is a vital piece of information for no-radio flying.

One important by-product of calculating the wind-correction angle is your expected groundspeed. Regardless of whether you correct for wind drift by drawing a wind vector on a piece of paper, the back side of an E-6B, or whether you use an aviation electronic calculator to get the answers, the solution to the wind problem will also give you your predicted groundspeed.

You've got to know an estimated groundspeed in order to then calculate the entries in your navigation log's columns for the time between respective checkpoints. Use whatever computer you prefer to do the math; just be sure that you do it. If you don't know when to expect to see a checkpoint come into view, you'll be lost in a short time.

You are not done after applying the wind-correction angle to figure your true heading from the true course of 104°, as measured with the plotter off the chart and the course line that you drew. You need to look at the chart again and see what the magnetic variation is in the area. Near your first checkpoint, the magnetic variation is 5° west. Just west of Parkersburg the variation line on the chart shows 6° west, and near Weston it's 7° west. This is enough variation to be concerned about correcting, so average it out at 6° west, and add 6° to your true heading to determine the magnetic heading. Remember west is best (+), and east is least (−).

Look at the compass correction card on the instrument panel of the airplane. If the card is missing, stop right now and either put the trip off or get the mechanic to "swing the compass" and record the findings. Apply whatever average correction is needed to accommodate the mechanical error in the compass, and you finally have the compass heading that you'll fly to make this trip. The heading will be fine-tuned as necessary en route by observing your checkpoints and making whatever heading corrections are needed to fly the intended path over the ground.

## NAVIGATION LOG

Refer to chapter 4 and the comments about a navigation log. Using a log is imperative for a no-radio cross-country. You must carefully calculate the time that it will take to get from one checkpoint to the next; total flight time is used for fuel planning and the VFR flight plan. If you don't go through this step, it's almost a sure thing that you'll get lost sometime flying these trips.

Keep in mind that one reason why a safe pilot occasionally flies a no-radio cross-country is to keep in practice for the time when the electrical system or radios fail and there is no other way to complete the flight. If you don't have a navigation log filled out and in use, how can it help you complete a routine electronic-navigation cross-country when the black boxes call it a day? See Fig. 5-3 for a completed navigation log for this flight, which assumes a no-wind situation solely for the purpose of illustration.

Dead reckoning still has a place in the age of VOR, loran, and GPS. It's not just a quasi-emergency procedure if the radios quit; it is simultaneously fun and a challenge. Most pilots appreciate these aspects of flying, and you likely will too when you become proficient at it.

If you're a renter pilot who longs to own a personal airplane but can't justify the cost of a modern, well-equipped aircraft, don't forget the classics. In 1994 dollars, $15,000 to $20,000, or maybe even a little less, can still buy a lot of capability for fun flying. A few airplanes with good cross-country speed and load carrying capabilities, like a Piper Tri-Pacer, can still be had in that price range.

Many two-place airplanes, such as the Cessna 120 or 140, offer quite a bang for the buck if you're willing to forgo all of the fancy boxes and get out your chart, plotter, computer, and navigation log and fly the way that pilots did for decades before the black boxes became available to general aviation aircraft.

Be careful learning this way of navigation because it might be unfamiliar to pilots trained in the last 20 years or so, but once practiced and mastered and mixed with a large dose of good judgment, it can be as safe and satisfying as it was before the advent of the microchip.

| NAVIGATION LOG | | | | | | |
|---|---|---|---|---|---|---|
| ROUTE OF FLIGHT: *FAYETTE COUNTY (CSS) DIRECT TO ELKINS - RANDOLPH COUNTY (EKN)* | | | | | | |

| Checkpoints | Altitude | Mag. Hdg. | Fuel | Dist. | GS | Time Off |
|---|---|---|---|---|---|---|
| | | | Leg | Leg | Est. 119 KTS | |
| | | | Rem. | Rem. | Act. | ETE |
| | | | 8 GPH 40 GAL | 171 NM | | ATE |
| ABEAM PICKAWAY Co. + ROSS Co. AIRPORTS | 7,500 | 109 | 1.20 | 19 NM | | 9 MIN |
| | | | 38.80 | 152 NM | | |
| ABEAM ADELPHI | 7,500 | 109 | 1.00 | 13 NM | | 7 MIN |
| | | | 37.80 | 139 NM | | |
| ABEAM VINTON COUNTY AIRPORT | 7,500 | 109 | 1.00 | 16 NM | | 8 MIN |
| | | | 36.80 | 123 NM | | |
| ATHENS, OHIO | 7,500 | 109 | 1.00 | 14 NM | | 7 MIN |
| | | | 35.80 | 109 NM | | |
| CROSSING OHIO RIVER | 7,500 | 109 | 1.36 | 20 NM | | 10 MIN |
| | | | 34.44 | 89 NM | | |
| CROSSING I-77 | 7,500 | 109 | .56 | 8 NM | | 4 MIN |
| | | | 33.88 | 81 NM | | |
| ABEAM HARRISVILLE | 7,500 | 109 | 1.44 | 22 NM | | 11 MIN |
| | | | 32.44 | 59 NM | | |

**Fig. 5-3.** *The first page of the completed navigation log for the cross-country flight from CSS to RQY.*

# 6
# Using avionics

NOW WE'LL LEAVE THE WORLD OF BASIC CROSS-COUNTRY FLYING WITHOUT electronic aids to navigation and venture into a realm that is more the norm for pilots flying modern lightplanes today. Other than the basic reliability that has been designed and manufactured into today's light airplane, nothing has made general aviation a safer, more pleasant, and efficient way to travel than has the astounding revolution in avionics that we have seen since the early 1970s.

Just those few short years ago, most lightplanes had only VOR receivers as the primary means of en route navigation, perhaps with an ADF onboard if the airplane were used for instrument flight. Many aircraft did not have altitude encoding transponders. Loran receivers were just barely coming on the scene, but they still were not in very many planes. GPS was unheard of.

I recently learned from a friend that a flight of Air National Guard A-7s flew from Ohio to Turkey with one of the pilots using a general aviation hand-held GPS receiver as a means of over-ocean navigation. The GPS unit was as accurate as the expensive military hardware onboard the fighter. What the government once had to itself, at tremendous cost, is now available and affordable to everyone. While it is beyond our scope here to delve into the electrical engineering and innards of avionics, we will examine the practical use of them from the perspective of the pilot in the cockpit.

## THE BASIC PANEL

The flight instruments that are installed in a modern airplane aren't described within a correct definition of avionics, but modern cross-country flight wouldn't be possible without them. The kinds of airplanes that we flew back in the 1940s, '50s, and even into the '60s had only a magnetic compass, and sometimes a venturi-driven turn-and-bank indicator to aid navigation. My old Taylorcraft had no gyro instruments at all, not even the turn-and-bank. You held a heading with the "whiskey" compass and kept the airplane level and upright by looking out of the windows with no attitude indicator or other modern instrument in the panel for reference.

Today's well-equipped lightplane has a panel full of instruments that would be equal to all of the airliners of just two or three decades ago and not that much different from the newest ones, except that the most recent generation of airline and exotic turbine-powered business aircraft are now coming out with a total electronic instrument panel that is computer-driven and displayed. These electronic flight information system (EFIS) panels are collections of computer screens upon which the system paints the look of an artificial horizon overlain with weather radar and navigational information.

Some EFIS display screens scroll through checklists for routine and emergency procedures and to annunciate various events and anomalies to the flight crew. Many EFIS units paint the weather-radar return right on the face of the horizontal situation indicator image, and the position of the airplane relative to the weather return is instantly observable.

We'll assume that your airplane is equipped with a full complement of mechanical, as opposed to computer-generated, gyro-driven flight instruments: attitude indicator, turn-and-bank indicator or turn coordinator, and directional gyro. In addition, your airplane has modern pressure instruments: airspeed indicator, altimeter, and vertical speed indicator. Except for the vertical speed indicator (VSI), the pressure instruments have been around for a long time and are required by the FARs for even the most basic airplane.

## NAVIGATION AND COMMUNICATION RADIOS

Radios with which you can navigate and by which you can communicate come in a profusion of models. The basic function of these is the same: to receive signals from some kind of transmitter and display the information in a meaningful way to the pilot to aid in navigation and enable the pilot to communicate with someone on the ground or once in a while with the pilot of another aircraft (Fig. 6-1).

Most of these units are called *navcoms*, which combine a VOR receiver and a communications transmitter and receiver together in one box. They're called one-and-a-half units because they consist of a full communications send and receive capability, and a receiver only for navigation. As this book is being written, a new kind of navcom has come on the scene, which combines the communications radio with a GPS receiver, eliminating the VOR function altogether. But for now, we're going to assume that a navcom is of the standard variety: communications and VOR reception.

**Fig. 6-1.** *Typical navcom radio.*

The communication radio transmits on a number of channels. The old units actually had crystals in them to tune the transmitter to the selected frequency. The newer units do the tuning electronically and have become more maintenance-free because the mechanical tuners that used to wear out have been eliminated. The standard number of communication channels is 760, but a radio with 720 channels will still get the job done in all but the most congested places.

Don't buy a used navcom that has only 360 channels because that is not enough for today's airspace and frequency congestion. Even if you don't fly in busy airspace, the frequency stability and tolerance of these ancient 360-channel sets don't meet the modern requirements of the Federal Communications Commission. A few years ago the FCC tried to outlaw the use of many of the 360-channel radios still in common use. The FCC rescinded that rule, but it could always be reissued, which would require the removal of any units affected.

Even if no such rule is ever promulgated again, parts for these old sets are getting very hard to find because the original manufacturers have for the most part quit supporting them with parts and service because the units haven't been manufactured for more than two decades. For sure, if you only have 360 communication channels available, someday you'll go someplace where you won't be able to talk with the controller on the primary frequency; therefore, make sure that you buy at least a 720-channel navcom.

In some of the more expensive models of radios, the com and nav functions are not combined in the same unit, but are separate. These installations have separate control heads and displays mounted in the instrument panel for both kinds of information, and the guts of the radios are remotely mounted somewhere else, often in the nose if it's a twin-engine airplane (Figs. 6-2 and 6-3).

## OMNI BEARING INDICATOR/SELECTOR (OBS)

The OBS has been around since the advent of the VOR system in the 1950s, when VORs came into general use, replacing the low-frequency radio ranges that had been

**Fig. 6-2.** *Digital communications transceiver.*

**Fig. 6-3.** *Digital navigation receiver.*

the foundation of radio navigation since the 1930s. In the first-generation VOR radios, the OBS was usually put right in the radio unit itself. Since the early 1960s, the OBS has been customarily remotely mounted away from the radio control unit.

A typical OBS is a circular dial with the degrees of the compass rose around its outer edge. The compass rose rotates by means of a small knob so the pilot can select a specific course or radial on which to fly. In a few examples of the OBS, the compass rose might be fixed in place, and the knob might rotate a pointer to the direction that the pilot selects.

The OBS has a vertical needle that oscillates from right to left to indicate a relative amount of deviation in the present position of the aircraft from the course or radial selected on the rotating dial. This vertical needle is called the *course deviation indicator* (CDI).

The last part of the OBS is a TO/FROM indicator that is used to establish aircraft position on the selected radial and tell the pilot whether the aircraft is presently on the TO or FROM side of the VOR transmitting station with respect to the selected VOR radial.

When using the system properly for en route navigation, the pilot needs only to correct the airplane's heading toward the deflection of the vertical needle to intercept the selected bearing or radial and then keep the needle centered to track that bearing or radial. Some OBSs use a bar of horizontal lights to accomplish the task of telling the pilot whether there is any present deviation in the aircraft's position from what is selected. These light-bar CDIs aren't very numerous, but do seem to work well.

Whether your airplane is equipped with a needle or light-bar CDI, either one is used in the same manner. The common mistake made by all too many pilots involves incorrect technique for correcting deviations from the course dialed into the OBS. This instrument only shows position. It displays the present position of the airplane relative to the inbound course or the outbound radial selected. (Radials are always numbered by their magnetic course outbound from the station; bearing is the magnetic course inbound toward the station).

Assume that you're flying toward a station with 270° selected in the OBS, and assume that the station is truly west of your present position, so that you have to fly west to get to the station. Those who are schooled in VOR use know that you're actually tracking inbound to the station on the 090° radial because radials are named by their outbound magnetic courses from the center of the station.

Now let's further assume that the needle in the OBS is deflected about two dots to the right, or if you have the light-bar type, two lights are illuminated. What's this telling you? All that it says is that you are positioned to the left of course, and the course that you've dialed into the OBS is off to your right, but not by much. You've got to remember that a VOR receiver and CDI do nothing more than tell you your present position relative to the bearing or radial that is dialed into the OBS.

How do you get back on course? That depends. Only one thing is for sure; you don't arbitrarily start a right turn and continue it. Remember, your present position is slightly to the left of the desired course, so the course is off to the right. How far to turn and when to stop the turn take some analysis and practice. Many people think you just fly toward the needle. That's true, but it doesn't tell the whole story. What you really need to do is to reintercept your desired course that lies off to your right.

The radials of a VOR project outward from the station like the spokes of a wheel. The closer you are to the station (the "hub" of the wheel), the closer together are the radials or the spokes. When the needle has a two-dot deviation and you're within 5 miles of the station, it's no problem because you are very close to your desired course; five miles from the station the radials are very close together. Most VFR pilots cannot keep a needle centered as they pass within 5 miles of the station. IFR pilots are trained and expected to do so, but most of them will see some wiggling of the course deviation needle close to the station. Fifty miles out the radials are much farther apart. To be able to know how much correction to make involves knowledge of how far you are from the station and how strong any crosswind component is causing you to drift.

If you're a long way out from the station, and if the wind is blowing hard enough to force you off course to the left, your correction has to compensate for both of the facts that you've got to first overcome that wind drift and then turn somewhat farther to the right to get back to the selected course. If you've drifted off course through inattention without the wind's having blown you off, you only need enough "cut at it" to reintercept the course.

The point is that you turn some and then stop the turn, taking studied angular cuts to get back on course. Turn some, usually no more than about 30°, and see what happens to the CDI needle. If it slowly starts creeping back toward the center, after 2 or 3 minutes, hold what you've got until it's centered, then remove ½ of the correction angle. If you stay on course, great; if you wander a little to either side, take a very small cut toward the needle, hold that heading, and eventually you'll discover the heading that will keep the needle centered. The trick: Don't overcorrect and zigzag back and forth across your desired course.

Depending on the VOR station being used and also depending on whether there are any terrain features blocking or interfering with the VOR signal, you might experience "scalloping" while navigating with VOR. Scalloping is the term used to describe little wiggles that you might see in the needle of the CDI. If they occur once in a while, that's normal. But if the needle is never stable, it's time to see the radio shop.

## TRANSPONDERS

Transponders evolved from the "identification—friend or foe" device that was initially developed during World War II that enabled ground-based controllers in the days of radar's infancy to tell the good guys from the bad. The modern transponder receives interrogation signals from radar on the ground and automatically responds to it. In this way, the image on the radar controller's screen is enhanced with a symbol that is different from what is seen from just the normal reflection of the radar's primary signal, and each transponder equipped aircraft is easily seen, identified, and distinguished from the other ones that appear on the controller's scope (Fig. 6-4).

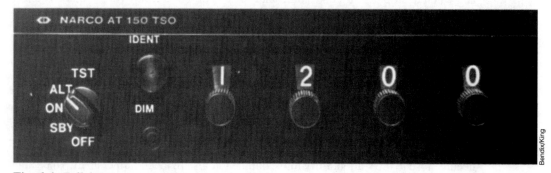

**Fig. 6-4.** *Full-feature transponder.*

The FARs now require the use of a transponder whenever flying in any controlled airspace if the airplane is equipped with a unit that has undergone its biennial recertification. If you're there, you must turn it on. Additionally the rules mandate the installation and use of Mode-C transponder equipment to operate in Class A, B, or C airspace.

Mode-C is the acronym for altitude-encoding capability on the part of the transponder. The transponder has another device, called an encoder, wired into it so that the altitude of the aircraft can also automatically be reported to the radar computer and displayed on the scope beside the airplane's target symbol that the controller sees.

The encoder might take its information from the altimeter, in which case it's called an *encoding altimeter*, or it might be a separate box that is called a *blind encoder*. In either type, the encoder is set to 29.92" Hg, so it is telling the transponder the pressure altitude of the airplane. The radar receiver on the ground has the software built into its system that enables the computer to convert the raw pressure altitude data, using the current station's barometric pressure, into usable altitude information. Just like the responses to the interrogations from the ground-based radar, the transponder does all of this automatically once the pilot has input the transponder code assigned by the controller (Fig. 6-5).

KEA 129 Encoding altimeter to 20,000 feet.  TSO'd.

KEA 130 Encoding altimeter to 35,000 feet.  TSO'd.

**Fig. 6-5.** *Encoding altimeters*

When you're using Mode-C, the controller will first ask you to confirm the encoded altitude as a cross-check against spurious data. One time when I was flying my Comanche on a cross-country of only about 100 miles, the controller asked for my type of aircraft, to which I responded that I was in a PA-24. He then asked if it was a turbo Comanche, and I said, "No." I was told to turn off the Mode-C because it was transmitting 17,500 feet and climbing steadily. Mode-C works most of the time, but always confirm its output with the controller first before relying on it.

A transponder has become required equipment for any airplane that is going to be used for serious cross-country flight. You can't go into any Class B or C airspace without one. A good number of very well-equipped aircraft even have dual transponders installed so that the aircraft's operational limitations aren't affected if one unit fails.

Most single-engine lightplanes can get by with one transponder. Plan to immediately buy one if you acquire an airplane that doesn't already have a transponder in its panel. If you're considering the purchase of an airplane and notice that the subject of your heart's desire has a transponder, be sure that you ascertain that it is a Mode-C unit with an encoder. For years transponders were installed without encoders, but the modern rules require Mode-C capability, and you could be cheating yourself if you get an older box that doesn't encode altitude.

## GLIDE SLOPE AND MARKER BEACON RECEIVERS

These avionics are normally reserved for the IFR pilot and are rarely used during VFR flight. The glide slope is a part of the instrument landing system (ILS), and its data is processed by a separate receiver, and displayed on the OBS by a horizontal needle, or set of vertical light bars. When the VOR receiver is tuned to a localizer frequency, the glide slope frequency is automatically tuned with the localizer because the two are paired together in a worldwide standard.

The horizontal needle or vertical light bars tell the IFR pilot whether she's above or below the glide slope of an instrument landing system. The needle that is normally used for VOR navigation tracks the localizer signal and tells her the aircraft's alignment with the runway centerline. The glideslope needle is flown in much the same way as is the course deviation needle. If the horizontal needle is deflected upward, above the center circle on the OBS face, it's telling the pilot that the airplane is below the glideslope, so the pilot needs to reduce the rate of descent in order to reintercept the glideslope. Conversely, if the needle is below the center circle, the aircraft is above the glideslope, and the rate of descent needs to be increased to get back on the glideslope.

Marker beacons are another part of the ILS, and there are typically two. These tell the IFR pilot the distance from the end of the runway. The beacon transmitters activate a respective flashing light and/or aural signal in the receiver. As the aircraft flies over the outer and middle markers (there are not many inner markers in use anymore), the light flashes or the buzzer sounds, and the pilot knows the airplane's progress along the localizer during the ILS approach.

## AUDIO PANELS

These devices are an array of switches that enable the pilot to select which of the various other avionics are to be channeled to the cabin speaker, through a headset, or neither. Because most modern lightplanes have two communications radios, the audio panel also has a switch to determine which transmitter will be fed signals from the microphone. Many audio panels also incorporate an amplifier to enhance the audio output from your other radios, and most also have built-in marker beacon receivers and display lights (Fig. 6-6).

Many modern communications radios do not have a built-in audio amplifier and depend upon an audio panel to provide the needed amplification of the audio signal to

KMA 24 Audio control console (actual size)

KMA 24 (International) with HF and Auto (actual size)

**Fig. 6-6.** *Most modern audio panels include an audible marker beacon receiver and visual beacon lights that are part of a localizer instrument approach.*

drive the cabin speaker. If your airplane doesn't have an audio panel installed, be certain that any radio that you buy already has a built-in amplifier, or you'll have to also purchase an audio panel to make the new radio work properly.

An audio panel greatly simplifies the job of selecting which of the radios will feed into the cabin speaker or your headset at any given time. Because most modern aircraft are equipped with two navcoms, most pilots will set one radio up to talk to the ground controller and tune in the tower frequency on the other. In that manner, a flip of a single switch on the audio panel changes from transmitting and listening on one radio to the other radio. In the same way, you can set the departure control frequency into the radio that had been tuned to ground control so that immediately after takeoff all you have to do is switch the audio panel over to select the navcom that is tuned to departure control.

Without an audio panel there's a whole lot more juggling of switches needed to tune in new frequencies when you fly in busy airspace where you're constantly being transferred from one controller to another. Consider the purchase of an audio panel a necessity if you regularly fly out of or into high-density areas.

## AUTOMATIC DIRECTION FINDER (ADF)

The ADF is one of the oldest radio navigation aids and was in general use before World War II. It's still valuable primarily for the IFR pilot rather than the VFR pilot. Recall that navcoms and VOR receivers work in the very high frequency (VHF) bands; the

ADF receives signals that are in the AM band slightly below the commercial AM frequency spectrum.

Most of the units can be tuned to frequencies from 190 MHz to 1,799 MHz; therefore, they receive not only dedicated nondirectional beacons (NDBs), but also the standard AM broadcasts. They consist of a tuner, instrument face that looks like a compass rose, and speaker output. Figure 6-7 is a modern ADF receiver.

**Fig. 6-7.** *Automatic direction finder (ADF).*

When an ADF is tuned to a station, the needle in its round readout just points directly to the station. If it's pointing straight ahead, the station is directly in front of you. If the needle's head is pointing to your left, that's where the station is and in like manner on around the face of the readout. Turn the aircraft until the needle points straight ahead, and you're flying right toward the station.

The problem with this system is that the readout doesn't allow for "needle-simplified" wind correction like a VOR receiver. If you've got a crosswind on your course, you have to learn some mental mathematical gymnastics to track into the station in a straight line. That station won't remain directly in front of you when there is a crosswind blowing at an angle to the route of flight unless you apply a wind-correction crab angle.

The quality of ADF reception is dependent upon local atmospheric conditions, and during stormy weather the electrical interference from the weather can reduce an ADF to nothing more than a good paperweight because the speaker or headset crackles, and the readout wanders aimlessly due to the electrical energy produced by the storms nearby. On a good day, especially at night, ADF signals can be received for hundreds of miles.

One of the nicest uses of ADF for the VFR pilot is the ability to use a regular AM radio station for navigation (or to listen to the ballgame). If you're headed for a smaller city or town that is not close to a VOR station, you can home into its AM radio station, if you know the frequency. Don't try this for long-range navigation until you get some instruction in the tracking of an ADF course because you are probably very used to tracking VOR radials with the OBS needle centered to correct for winds.

In tracking an ADF course, you've got to correct for the wind drift yourself. It's not particularly hard to do; you just have to spend a few minutes of ground discussion time with an instructor, preferably an instrument instructor, and then go out and fly for a while to get the hang of it.

Because the continental United States has a VOR system that is very complete in its coverage, ADF is not normally used for en route navigation. But if you venture into a good deal of Canada or Mexico, take some instruction on the use of ADF for long-range navigation because in those countries, as well as most of the third world, it's all you've got.

If you don't know how to do the mental math involved in ADF tracking, don't try to teach yourself on a cross-country flight in one of those places, or you might end up a hundred miles off of the direct line you want to fly and run out of gas before you can home in on the station.

## RADIO MAGNETIC INDICATOR (RMI)

The RMI has been around as long as VORs, but its cost and complexity prevented popular use in lightplane flying. Until recently, it was heavily used by the military. It has a circular readout much like an ADF, but has two needles. The RMI accepts input from the VOR and ADF receivers. The fat needle points to the NDB to which the ADF is tuned; the thin needle points to the selected VOR station.

Most RMIs have a rotating compass card that is automatically slaved; thus, it also shows the pilot the magnetic heading of the aircraft in the same manner as does the directional gyro, but all in one instrument. An RMI is confusing until you learn to use it, but don't worry because it's doubtful you'll ever see one in a lightplane.

## DISTANCE MEASURING EQUIPMENT (DME)

DME does just what its name implies; it tells you your distance from either a VORTAC or a VOR with DME capability. A VORTAC is a VOR station, collocated with a TACAN, which is the military system of VHF navigation. The TACAN receives a signal transmitted from the airplane's DME and returns it to the unit in the panel. The DME measures the time that it takes for the round-trip and calculates the distance to the station. After a few successive readings, which take place in seconds, the DME can calculate the airplane's groundspeed in knots.

Most modern DMEs have tuners that are channeled together with the VOR receiver because the DME actually operates in the military UHF band so that there is no need to separately tune the newer DME units (Fig. 6-8). If you encounter an older DME with a separate tuner, you still tune it to the VOR frequency, and the pairing function sets it to the proper TACAN channel. A switch usually enables the pilot to select which VOR receiver (assuming two onboard) will also automatically tune the DME to the VOR station.

A DME is telling you the distance from its antenna, which is normally mounted on the bottom of the aircraft, to the TACAN station to which it is tuned. This distance is called *slant range* because there is no altitude correction in the readings that the pilot sees in the cockpit. If you're flying at 10,000 feet AGL, and pass directly over the station, the DME will indicate 2 miles as the closest that you'll ever be to the station. At

KN 62A DME    Distance/groundspeed/TTS (GS/T mode)

KN 62A DME    Distance/frequency (Freq mode)

**Fig. 6-8.** *Distance measuring equipment (DME).*

5,000 feet, it would show 1 mile. This "error" between the real distance and the slant-range to a station isn't meaningful until you're close to the station. Any published DME distances that you see on IFR charts already have the slant-range difference built into the published figures.

DME is still a part of the standard airway and ATC system in the United States, and if you're going to be doing any real IFR flying, your airplane must have one. Unfortunately, they're expensive and might soon be supplanted by the data calculated by a GPS unit.

## AUTOPILOT

An autopilot is a great convenience for any pilot who flies much cross-country. It can range from a simple wing leveler to a sophisticated 3-axis model that can couple ILS approaches, often reducing the pilot to a systems manager. Most lightplane autopilots are wing-levelers with a heading-hold function.

These units allow the pilot to be relieved of constantly working the ailerons to keep the wings level and the airplane on heading, which relieves a great deal of stress during cross-country flights. Units that do these two jobs are affordable for most aircraft owners who want a capable cross-country airplane.

KG 258

Artificial horizon

KAP 100

Single-axis autopilot with panel-mounted
computer, mode controller and annunicator.
Manual electric trim option available.

KC 190

Digital computer, mode controller, annunicator

KG 107

Directional gyro

**Fig. 6-9.** *Single-axis autopilot.*

Figure 6-9 shows a typical single-axis autopilot with heading and nav hold functions plus the two gyro instruments that feed the necessary operational data to the autopilot. You are more likely to see this configuration in a lightplane installation.

The more exotic autopilots that can capture and hold an altitude, track VOR radials, and even couple and fly instrument approaches are quite expensive, and are usually found only on high-performance singles and twins. For single-pilot IFR, some type of autopilot is a virtual necessity. For VFR work, you can get along without one, but once you use one on a long trip, you won't want to be without it again. Simple jobs like refolding a chart or looking up some information in the airplane's handbook don't require juggling between keeping the wings level and doing the task at hand when you have even the simplest autopilot installed.

Today we have some autopilots that have become so capable that they are better called *integrated flight control systems*. The ones on the newest generation of airliners even manage the power settings after computing all of the atmospheric and aircraft performance and weight factors to arrive at that one level of power that is most economical. Some can even land the airplane hands-off with the pilot only watching. You won't find any of these in a lightplane because the system costs many times the value of such a small airplane.

## FLIGHT DIRECTOR

Contrary to a popular misconception, a flight director is not an autopilot, although it's generally integrated with one. A flight director is a computer that simultaneously accepts information from all navigation and air-data instruments and combines this data into a single readout or pictorial display that tells the pilot what action to take with the controls to achieve a desired result.

When coupled to an autopilot, the flight director tells the autopilot what to do, but the flight director can be used without the autopilot's being turned on when the human pilot follows its lead. In most installations, the autopilot can also be used with the flight director turned off. Flight directors have become more common on high-performance singles and light twins, and are almost always found in turbine-powered aircraft.

## HORIZONTAL SITUATION INDICATOR (HSI)

An HSI is a highly sophisticated piece of equipment that combines the information usually gathered from a directional gyro, VOR OBS, CDI, localizer, and glide slope all into one cockpit display. HSIs used to be quite expensive but have been dropping in price over the last decade to the point where they are often installed in high-performance singles and light twins. They are usually a part of the installation of a fully capable autopilot.

I first used an HSI in the early 1970s when they were uncommon in lightplanes; our company had a Piper Aztec. When you get used to an HSI, you'll wonder how you ever flew instruments without one. They are a nice plum for VFR pilots but have almost become "required" equipment for an IFR airplane (Fig. 6-10).

An HSI combines so much information into one instrument that the use of one greatly reduces pilot workload. Instead of having to gather the data concerning the airplane's heading and navigation from two or more separate instruments, an HSI displays all that at once. If you buy an airplane intending to redo the avionics stack or update the radios in an airplane that you already own, seriously consider adding an HSI to your panel; you won't regret it.

## AIRBORNE WEATHER RADAR

The sole function of airborne weather radar is to "see" areas of precipitation. When flying in fronts or other weather systems that have the potential to spawn thunder-

**Fig. 6-10.** *Horizontal situation indicator (HSI).*

Bendix/King

storms, the intensity of the turbulence is often related to the corresponding intensity of precipitation. Radar is not of particular value to the VFR pilot because he shouldn't be in the clouds in the first place. But the IFR pilot flying a radar-equipped airplane can "see" areas that might be bumpy when he can't see outside the airplane's windshield.

It takes a lot of practice and schooling to learn to interpret the returns seen on the radarscope, especially on the older monochrome black-and-white units. The modern color radars automatically grade the intensity of the precip into colors from green through red, with red being the heaviest, and therefore likely to produce dangerous turbulence.

One problem with the use of radar is the phenomenon of *attenuation*, which means that the radar signal is absorbed and decreased by the precip that it's showing. When that happens, you won't see threatening weather that is lurking behind a cell of precip being displayed. That's just one of the reasons why you have to learn how to use and interpret radar data.

There are some less costly radars that have found their way into high-performance lightplanes, but radar is still a rarity in singles. Several radars are on the market that can be installed in a pod beneath the wing or in the leading edge of the wing if you are willing to go to this expense in a single-engine airplane.

## STORMSCOPE

The Stormscope was invented in central Ohio by Paul Ryan. He had taken his family on a flying vacation when they got bounced around by an inadvertent encounter with a thunderstorm. At the time, no one had yet figured out how to install weather radars in single-engine airplanes because the antenna of the radar always went in the nose cone of multiengine airplanes. Ryan invented a device that locates thunderstorms by sensing the electrical discharges of the lightning bolts and displaying that information on a circular instrument in the cockpit.

The rights to the Stormscope were subsequently sold by Ryan, and it has gone through several stages of evolution since its beginnings. Some pilots prefer it to radar because radar only sees precip, and you can have an active area of turbulence that isn't producing rain. Another advantage to a Stormscope is that it can display information in the full 360° circle around the airplane.

Radar can only show what's within about 30° or 45° on either side of straight ahead. Stormscopes are somewhat less expensive than radar and are a much simpler device to install, use, and maintain. Ryan's invention has been a tremendous aid to general aviation pilots who like to stay out of thunderstorms (is there any other kind?), and he is a good friend and fellow aviator.

A Stormscope can even be used on the ground before beginning a flight to see what electrical disturbances are around. You should never turn on a conventional radar before takeoff because its microwave energy can be very dangerous for anything in front of the airplane, including people.

## TRAFFIC COLLISION AVOIDANCE DEVICE (TCAD)

TCAD is another product of the genius of Paul Ryan. Since selling the rights to the Stormscope, Ryan has developed a relatively inexpensive collision avoidance device.

Several years ago, the FAA mandated a schedule of future installation dates for a traffic collision avoidance system (TCAS) in airliners. TCAS is a very capable system that senses the presence of other aircraft that pose a collision hazard. Modern versions give the pilot commands of whether to climb, dive, or turn to avoid the threat. False alarms are inevitable, but they are limited. The main problem with TCAS is the extraordinary expense that puts it beyond the means of almost all general aviation aircraft owners.

TCAD is a simpler device that senses the presence of other aircraft equipped with and using a Mode-C transponder. In essence, TCAD receives the transponder signals and tells the pilot where the threat aircraft is but doesn't give avoidance commands. You have to interpret the data given and decide for yourself how to resolve the threat. The downside of TCAD is that it senses only the presence of aircraft that are equipped with and using an altitude-encoding transponder.

If you fly into or out of high-density traffic areas with much regularity, think about investing in a TCAD. I know only one person who survived a midair collision, and she was flying a glider. Perhaps the relatively slow speeds of the two colliding gliders saved her life. Enough of the one damaged wing remained on the aircraft that it descended to the ground like a falling leaf, which tremendously softened the ensuing impact, rather than come down with more vertical speed. She still flies and instructs in gliders.

## HIGH-FREQUENCY RADIO (HF)

HF transceivers operate in the high frequency bands where their range can extend several thousand miles. Before VHF frequencies came into use after World War II, all aviation communication was via HF. Today HF radios are relegated primarily to overwater use where their range is vital and outweighs the poor qualities of HF as a communications medium. They're almost never found in lightplanes. Portable HF units can be rented for overseas ferry missions.

HF radios generally require a trailing antenna up to 50 feet or longer in length, and the quality of the reception is dependent on atmospheric conditions. If you ever use one, be certain that the antenna is reeled back in before landing.

## AREA NAVIGATION SYSTEMS (RNAV)

Until the late 1970s, general aviation pilots were content to fly from one VOR to another with the occasional use of ADF for their electronic navigational aids. Loran-C was a bulky and expensive system for marine navigation of larger boats and ships. GPS was still the dream of some engineers. In those days, the only promising new system for general aviation pilots was RNAV.

RNAV, which is shorthand for random area navigation, makes it possible to navigate from one point to another instead of from VOR to VOR. It allows the pilot to create *waypoints*, which are phantom reference points to which she can steer. These waypoints are used by the pilot just as if they were VOR stations because the RNAV has electronically located a point in space to which it has "moved" the VOR. A pilot can put a waypoint anywhere within the simultaneous receiving range of two VOR-TACs because the system takes information from triangulating the signals from two VORTACs to electronically create a point in space (Fig. 6-11).

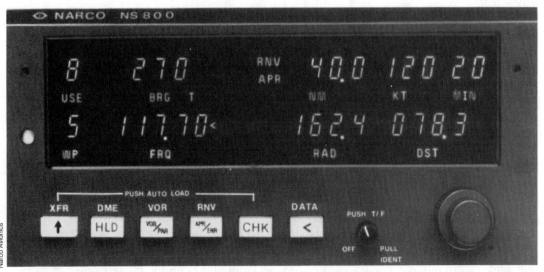

**Fig. 6-11.** *Area navigation unit, or RNAV, that uses VOR and DME information to enable the pilot to "move" a VOR station electronically.*

Older systems meet the same need to navigate point-to-point. VLF/Omega, inertial navigation systems (INS), and others have all been used by large and military airplanes. But all of them were either too heavy, too costly, or too hard for a single pilot to use to be of much use in a lightplane. Today RNAV typically refers to the VOR-TAC/DME-system of area navigation that we have just discussed.

Loran-C units have come down dramatically in price since the early 1980s. They have become an increasingly popular method of area navigation. Their accuracy can rival RNAV models, and they are within the price range of most lightplane owners. There are even hand-held loran units on the market in 1994. Unfortunately for those who invested in the higher priced lorans, they might soon by eclipsed by GPS, about which we'll say more in a short while.

Although GPS will likely take over in the next few years as the preferred method of point-to-point navigation, the traditional RNAV units still have some useful advantages.

Because RNAV can electronically relocate a VORTAC to a phantom location and enable the pilot to navigate to or from it just as if the waypoint were the true site of the VORTAC station, the use of RNAV is conventional for every pilot who has ever used a VOR. All that has to be learned is how to set up the "moving" of the station from its real location to an electronic waypoint. The airplane must have VOR and DME sets and the RNAV computer installed. You probably already have the VOR and DME, so adding RNAV isn't a big chore or expense.

The FAA has established RNAV routes between many airports, and most of the units are approved for use in IFR navigation. Quite a few RNAV instrument approaches already exist, so if you fly IFR, RNAV is a useful tool when you can't go VFR.

RNAV can be used everywhere that VORTAC coverage is not interrupted by any large geographical gaps. Like loran, RNAV might soon by relegated by GPS to the history of aviation, but that probably won't really occur until the turn of the century or later. If I were considering an avionics update to an airplane that I already own that does not have RNAV, I would think long and hard before going to the expense of adding an RNAV unit.

# LORAN

Loran is the acronym for "long-range navigation." There were earlier loran-A and -B systems that have now faded into antiquity. These early lorans were the size of a desk, and the entire system was originally developed for open-ocean marine navigation. The first generation units took a dedicated operator to figure out what was going on. The system never showed up in airplanes for all of these reasons.

The microchip changed all that, and smaller units are available that perform the trigonometry in a microsecond compared to the several minutes for the loran operator on a ship to calculate. Henceforth, the word loran will refer to loran-C, the modern system.

Loran came onto the aviation scene like gangbusters in the late 1970s and '80s. It did everything that RNAV did better, cheaper, easier, and with less pilot workload, but initially loran did lack IFR approval. Even though the electronic workings of the system are complex, modern computer chips make it easy for the pilot to operate.

Loran works by having transmitter chains that consist of master and slave stations. For the receiver to give useful information, it must be in range of a master and two slaves. It calculates position by measuring the differences in the time that it receives the signals from the three stations. The unit's memory knows where those stations are; therefore, some trigonometry occurs automatically. When all the data has been processed, the aircraft's present position is displayed according to latitude and longitude. Then it compares where it has been with where it is and tells the pilot the aircraft's course, groundspeed, time to destination, and wind drift (Figs. 6-12 and 6-13).

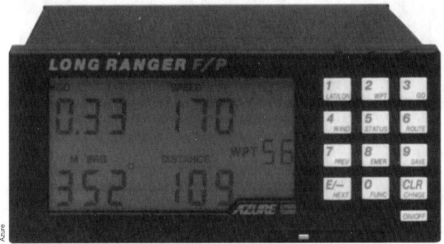

**Fig. 6-12.** *Loran unit that can be easily removed from the aircraft's panel.*

Two types of loran units are available: database in memory and no database. Some of the database units have replaceable credit-card-size memory cards that are updated every few months by the manufacturer, while the others with a database have a permanent memory. These databases know the location of airports, VORs, NDBs, airway intersections, and other sites of interest to pilots. If your unit doesn't have the replaceable card, you will probably have the ability to manually load points of interest to you, as long as you know the latitude/longitude coordinates of those places.

If you have one of the less-expensive units without any form of built-in database, you will have to input the coordinates that your trip requires, and then it can guide your way. You can get the information from a variety of sources, such as *Flight Guide*, *AOPA's Aviation USA*, and the *Airport/Facilities Directory*. Whether you've got one

**Fig. 6-13.** *A round loran receiver that was designed to fit in a standard 3-inch instrument hole in the panel. A built-in database warns the pilot before penetrating special use airspace.*

type of loran or the other, the effect is the same: point-to-point navigation without the zigzags involved in flying from one VOR to another.

At the airport before takeoff, you just ask the unit to tell you your present position, and you load that data in as the point of origination. Then input the coordinates of your destination and presto, in a few seconds the loran will compute the distance and course to fly between the two airports. After takeoff, it will compute groundspeed and the time remaining to fly to the destination. Most lorans display a course deviation indicator that is used just like the needle or light bars in a VOR unit, although the technical workings are much different.

If you want to know where you are at any time during the flight, the loran can almost instantly calculate your present position (Fig. 6-14). If you need to deviate from

IIMorrow

**Fig. 6-14.** *This blind altitude encoder enables the transponder to transmit aircraft altitude for display on a controller's radarscope. The encoder can also be connected to a loran unit to display altitude to the pilot.*

your planned route of flight and you don't have a database unit, all you need to know are the coordinates for the new place you want to go. If you have a more elaborate loran with a database, just dial in the three-letter identification of the new destination, and the new course will be figured before you can bat an eye.

During each flight with a loran as your means of electronic navigation, you'll fly as direct a route as possible, thereby using the least fuel, putting the fewest hours on your airplane, and saving your own time as well. Loran provides, next to GPS, the most efficient way to navigate over any meaningful distance and is far less workload intensive than RNAV to provide point-to-point navigation.

Another advantage of loran is its declining cost. We'll talk about GPS in a few moments, but with the advent of GPS as the newest of the black boxes available, the demand for new and used lorans has been declining; hence, so have the prices declined. In mid-1994, some lorans were advertised for as little as $400.

Most pilots who buy a loran want an automatic database and automatic triad ability. Because the loran must simultaneously receive reliable signals from one master and two slave stations, the automatic triad ability simply means that the unit finds and listens to the best combination of masters and slaves. If you buy an extremely inexpensive loran, you'll have to manually select the chain of stations.

One problem that might surface with the replaceable data card types is lack of manufacturer support. As GPS takes over, and I think there is no doubt that it will, you must wonder how long the manufacturers will continue to produce the replacement data cards. If they stop, you'll still be able to use the unit with the last card you received, but you'll have paid for an update capability that has become obsolete.

If you fly IFR, make sure that the box you buy has IFR approval, or you'll be back to VOR navigation in IFR weather. There was quite a bit of talk about establishing loran instrument approaches, but most of that movement has died down, prompted by the advent of GPS.

There are still lots of good loran units on the market, and some are even portable (Fig. 6-15). If you choose to buy one, shop around for the best unit at the best price. You can purchase a tremendous amount of capability in a loran unit for a few hundred dollars, particularly if you don't mind not having the latest technology in your panel.

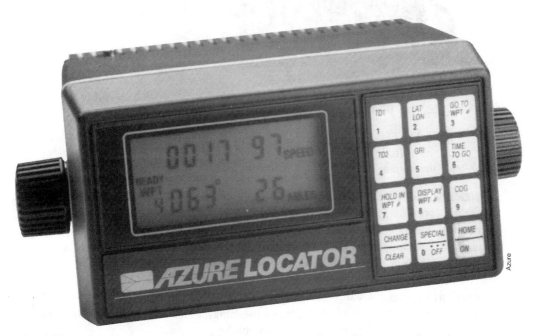

**Fig. 6-15.** *A portable battery-powered loran receiver can be used in any airplane, boat, or car, or by hikers.*

Because loran was originally developed for ocean navigation, the system was not required to produce reliable signals that could be received far inland, away from the ocean shores; hence, there was a midcontinent gap that resulted in a large area of the central United States in which loran could not be used. The government built more stations to deal with that problem, but if you live in the Great Plains area of the country, be sure that any unit you buy works reliably in your locale.

**106**

# GLOBAL POSITIONING SYSTEM (GPS)

GPS was known by the name Navstar early in its existence, but that label has fallen into disuse, and now virtually everyone in aviation refers to the system as GPS.

GPS was originally developed by the military to meet the aerial navigation needs of naval and military aircraft operating all over the world: within the United States, from carriers at sea, and in remote combat theaters far from any ground-based system of navigation. The other thing that motivated the invention of GPS is that it is totally independent of any type of transmitter on the ground because the military needs to be able to navigate during times of hostilities when ground-based transmitters could be destroyed by enemy action.

GPS is based upon a network of satellites in *geosynchronous orbit*, where each stays positioned above a given point on the Earth's surface, after being launched and established in that orbit. The satellite transmits signals that are received by the GPS receiver in the airplane. In order to work properly, the receiver must be receiving three satellites at the same time, and that's why 24 satellites make up the network. There is no place on the globe where you're out of touch with the minimum number of satellites, and that is why the system is called global positioning system.

From a pilot's perspective in the cockpit, a GPS unit is operated very much like a loran, and the similarity in controls of the sets on the civilian market was intentional to ease the transition from one to the other. In addition to all of the navigational functions such as present position, estimated time en route, course to fly, groundspeed, and the like, GPS offers one new bit of information never before available from an electronic navigation device: altitude.

The receiver is taking data from the satellites' output signals and measuring the time that it takes the signal to arrive from that fixed point in space, and from that time measurement the unit calculates how far it is from the satellite. Because it also knows the altitude of the satellite's orbit, simple arithmetic calculated within the receiver results in the receiver knowing how high it is above the Earth's surface, and displaying that to the pilot as the aircraft's altitude.

The accuracy of GPS is outstanding. In the military mode, it can tell position within a few feet, but the system's output is degraded somewhat for civilian use, down to about 100 yards. So with a GPS you'll know where you are anywhere in the world at any altitude within about 100 yards tolerance. Coverage is universal and reliable. Because a loran receiver depends on the ground-based chains of transmitters, there are places where the receiver will be out of range of the signals and therefore can't be reliably used. Not so for GPS.

The FAA and other governmental authorities are working on a new system of instrument approaches based upon GPS. Because we already have the basics of NDB, VOR, and ILS approaches in use all over the world, GPS approaches can be developed that simply overlie the profiles of the existing approaches. This is a much simpler task than was the goal of only a few years ago to develop loran approaches because they would have had to be established with new data from the loran transmitters.

With GPS approaches, all that really has to be done is to create new approach charts that depict the same approach profile as did the approach that was based upon the older ground-based NDBs, VORs, and ILSs. Because they have served us well for at least 50 years, the profiles and courses of the approaches have been proven. All we'll do is to fly the same directions, altitudes, and descent points using the GPS to guide us, rather than relying on guidance from ground-based transmitters.

GPS is the wave of the future. Because the military has already provided the satellite network, there's little more for civilian aviation to do other than produce and buy the airborne receivers. There might even be some future cost savings for the FAA if it is decided in the coming years to transition completely over to GPS and stop repairing and replacing the thousands of VORs and NDBs that dot the countryside.

GPS units come in either panel-mounted or portable varieties to suit the wishes of the customer. Most pilots who want to move over to GPS as their primary means of navigation will probably opt for a panel mounted unit. Don't discount the portable. Portables have a built-in antenna that is no taller than a soda-pop can, and a few have wire antennae that have to be strung along a window of the airplane. For the renter pilot, the portable is a real boon because it can be transferred in seconds from one airplane to another. These hand-helds sell for well under $1,000 and represent a great value. (At the time that you buy any GPS receiver, pay close attention to whether it is FAA-certified. A receiver that is not FAA-certified cannot be used for primary navigation, only for secondary navigational guidance. When the first GPS instrument approach was approved in 1994, only one brand of receiver was FAA certified.)

Portables are also useful for ground-based activities, such as boating and hiking in the wilderness. Some units are no bigger than a coffee mug and provide all of the functions of a panel-mounted GPS receiver as long as the batteries hold out. (As this book is written in mid-1994, some of the luxury auto makers are talking about mounting GPS units in cars and integrating them with a moving-map display to show highway maps right on the panel of your vehicle. Buck Rogers has come to life.)

One important advantage of the portable GPSs is that they provide state-of-the-art electronic navigation capability to pilots of older airplanes, like many of the classics that have no electrical systems. The trip in my old Taylorcraft from Ohio to Florida and back (chapter 5) would have been a cakewalk with a GPS receiver. Now pilots of these aircraft can navigate with the same ease and precision as can those who fly the best-equipped jets. Keep plenty of charged batteries available, and anyone can use a portable GPS.

## MOVING MAPS

Moving-map displays combine a map display with the output from either a loran or GPS receiver to show the pilot where the airplane is by reference to airways, airports, and ground-based navaids. This entire device was first available to general aviation in the late 1980s, and more of these units come onto the market every few months. Many GPS receivers come with a moving-map display as one of the several modes of operation that a pilot can select.

Moving maps also help you avoid areas of airspace that you might wish to avoid, such as Class B or C airspace, MOAs, restricted and prohibited areas, airports, and airways. If you have one, you won't have to worry any longer about whether you're about to fly where you shouldn't be. The moving map displays the aircraft's present position by means of either an airplane symbol or moving "bug" so that you can instantly see where you are in relation to the airspace that you want to fly around. If you want to use your loran or GPS to fly along a traditional VOR-based airway, the moving map enables you to do that, too.

## HAND-HELD TRANSCEIVERS

Hand-held transceivers look like walkie-talkies. Most of them also have the ability to receive VOR signals and can provide a navigational capability, albeit of lesser quality than the traditional panel-mounted unit due primarily to the limits of a hand-held's antenna. These units have also dropped in price over the last few years, and 1994 catalogues offered them for around $500 (Fig. 6-16).

I personally think that every serious cross-country pilot ought to invest in a hand-held radio, even if it's one without a VOR function. (Remember chapter 5; you're supposed to be able to navigate without any electronics.) They can be used to make initial calls to clearance delivery or to unicom for fuel or a cold-weather preheat and start. It will come in handy whenever you might need to make a radio call before the engine is running or after you've shut it down. If you have a hand-held, you won't need to turn on the airplane's master switch and run down the onboard battery.

Hand-held radios serve the obvious purpose of being a last-resort method of communicating if your aircraft's avionics or electrical system fails. If you fly IFR, they are a must piece of equipment. Even though the AIM has an entire section devoted to procedures to use if communications are lost while in IFR conditions, nobody with any sense wants to ever have to use them. With the advent of hand-held transceivers, nobody ever should use them, but the hand-held will never be a substitute for knowing the regulations and procedures.

In addition, they can be used at or in an office or at home for listening only. (Don't transmit from a ground location unless it is an emergency.) Monitoring various frequencies can help a new pilot learn radio-communication procedures in the comfort of a land-based chair. Also, a hand-held makes a wonderful emergency radio in case you are ever forced down. When the airplane battery runs down, or if the airplane is too damaged in the forced landing to be able to use its radios, you can get on the international emergency frequency of 121.5 MHz and get help. But if you're ever in this situation, don't expend the battery power of the hand-held making needless and fruitless calls in the blind. Make one or two calls every hour or so, then turn it off. When you do see a rescue aircraft, you'll want the battery power available to be able to talk then, when it counts.

## LONG-RANGE TANKS

Many light airplanes, back when they were manufactured in meaningful quantities, were offered with either standard or long-range fuel tanks. Many manufacturers' stan-

**Fig. 6-16.** *Hand-held transceiver with 760 communication, 200 navigation, and 20 memory channels. Every cross-country pilot should have one.*

dard tanks held a paltry supply of fuel, but with the reduced weight of the smaller fuel loads, the book numbers about useful and cabin loads look impressive, until you get to the disclosure about the range of the airplane.

Airplanes with a no-reserve range of less than 500 miles are somewhat impractical for serious cross-country flight and can even be an invitation to disaster. Fuel in the tanks means safety in many respects. A pilot with several hours of remaining fuel has many more options in dealing with the unexpected than does someone who is nearing the end of the road in terms of remaining gas.

If your airplane lacks sufficient fuel capacity for cross-country operations, you have a few remedies other than selling it and getting a new airplane. Ask your mechanic to look into the many aftermarket long-range tanks being offered by the modifiers. Most can be purchased and installed at your home shop without taking the airplane halfway across the country to the manufacturer of the tanks.

Most single-engine Cessnas can be modified with extra tanks that fit into and replace the standard wingtips. These tanks usually operate by gravity feed, allowing the extra fuel to be drained into the main tanks after the mains have been used for a while, which reduces the amount of fuel in them and makes room for more. Others have electrical transfer pumps to move the gas from the auxiliaries to the mains or directly into the engine.

Many other airplanes, such as Bonanzas, Comanches, Navions, and the like, can be equipped with fuselage or tip tanks that increase the available fuel supply. Whichever systems are approved for your airplane, they're a worthwhile investment for any pilot who uses an airplane for serious travel, especially IFR.

The fuel reserve and alternate airport requirements contained in the regs about IFR flying reduce many light singles to being nearly useless for IFR flying because the combined slower speeds and small fuel tanks render them incapable of flying to the destination, then to an alternate, then continuing to run at cruise power for the required time. Long-range tanks solve this problem.

If you're looking at competing installations, see if any of them come with an increase in allowable gross weight of the airplane. If they don't, your cabin load will suffer under the weight of the extra tanks and the fuel contained in them. Everything in aviation is a compromise, and the addition of auxiliary tanks often reduces the allowable load in the cabin and baggage compartments. The tanks themselves seldom weigh very much; therefore, when you need to carry a heavier load of people and luggage you just don't fill the auxiliary tanks. But when you can carry that extra fuel, it's well worth the investment to have the peace of mind that extra fuel gives you, whether VFR or IFR.

## ELECTRONIC FUEL TOTALIZERS

These gizmos are more than fancy fuel gauges because they constantly compute fuel consumption rates, fuel already consumed, and fuel remaining in the tanks. They work by having a flow meter very near the engine intake manifold; the meter very accurately measures the actual amount of fuel going through the engine.

The computer in the system can remember how much fuel was in the tanks at engine start and tell you what remains. The pilot usually updates the computer every time fuel is added, although some of the exotic units can sense that too. Most fuel totalizers also display the time remaining to dry tanks at the current rate of fuel consumption. You can immediately see exactly what effect a power reduction will have on the additional length of time you can remain aloft because the totalizer will quickly recompute the time remaining to dry tanks after you make a change in power setting.

Every student pilot should have burned into memory the fact that traditional aircraft fuel gauges are legendary for inaccuracy. That's why we teach primary students to time their fuel burn and not rely on the gauges to tell them anything upon which they stake their lives. Fuel totalizers have changed that by providing a pilot truly reliable and lifesaving information as long as any manual updating of the fuel supply is done properly when the airplane is refueled and as long as the unit doesn't break. That's why you should still keep track of your flying time, know how much fuel was in the tanks at takeoff, and monitor the rate of fuel consumption.

## AVIONICS CARE AND MAINTENANCE

Because avionics represent such a large investment, every pilot should be aware of what can be done to keep them running and reduce maintenance costs. Modern solid-state radios don't require the frequent repairs that the older tube-type units did, but they still do need occasional fixing.

The nemesis of all avionics is heat. Heat does more to reduce the reliability of radios and therefore increase maintenance costs than about anything else that you can control. If you tie down your airplane outside of a hangar, the interior of the cabin is going to get just as hot as the inside of your car when it is parked outside in the sun. Think of the multiple days that an airplane sits on the ramp between flights, and you'll quickly realize how much heat is working on your expensive black boxes for such a prolonged period of time.

Invest in one of the many cockpit window shields on the market. The shields are made of materials that reflect the sun's energy and keep the internal cockpit temperatures down quite a bit. Use these shields religiously and your avionics will last much longer and your trips to the radio shop will be fewer. As well, the interior of the airplane won't bleach and rot nearly as fast, and the glareshield will also stay nice looking much longer.

Look at the fuselage of your airplane on the side near the nose. You'll likely see a little air scoop that brings in a stream of air to cool the avionics stack. If your aircraft has one, make it a point during your preflight inspections to ensure that it's not blocked by dirt or wax. An electric fan behind the radios is there to cool them. If you notice that the fan isn't running when the electrical system master switch or avionics master switch is turned on, get it fixed. The radios will thank you with reduced repair bills.

There is a debate about whether to turn radios on that you aren't using. Many light-planes have an avionics master switch that turns on all of the radios at once even if you only need one of them to call a controller for taxi clearance. If your airplane has this convenience item, remember that most conveniences in life have a price tag attached. You can still use the individual radio on-off switches to decide if you want the avionics master switch to control all of them at once, rather than just turning on the avionics that you need to use at a given time.

Some radio technicians advise that solid-state radios aren't worn out by being on all of the time; other technicians think you should leave a radio turned off if you aren't using it. Who knows the real answer? I don't. Rely on a radio shop that you trust, and get its advice. The only thing about this that I do know is that the computers in our office are never turned off, and mine has been trouble-free for the four years since we got our current system; we always turn off our computer at home between sessions of using it.

# 7
# Engine and
# fuel management

TㅐE TYPICAL LIGHTPLANE HAS TWO SYSTEMS THAT REPRESENT THE costliest parts of the airplane. In the preceding chapter we talked about avionics; obviously all of the electronic aids to navigation and communications radios comprise a significant investment and make up quite a bit of the cost of any modern well-equipped light aircraft. The other system of major economic importance to the aircraft owner is the engine and, by corollary, the fuel burned by it.

We have already covered some of the things that a pilot can do to reduce the cost of maintaining the avionics in the airplane. Now let's take time to examine how some management techniques can be developed and used to keep the engine operating for as long as possible with minimal unscheduled maintenance and burn the least fuel consistent with safety and mission requirements.

## ENGINE STARTING

The first few seconds that an engine runs after starting generate some of the greatest wear and tear on it, especially if the engine hasn't been operated in the few hours before the start. We all know that the oil lubricates the engine and its internal moving parts. When an engine sits idle for just a few hours, the oil that coated the surfaces of the cylinders, pistons, bearings, and other parts that rub against each other starts to drain off the surfaces that abrade, and it returns to the oil tank or bottom of the crankcase.

When the engine is started after the oil has drained off of those surfaces, it is running without benefit of lubrication for a short while. Abusive starting can exponentially increase the wear that the engine suffers during these vital few seconds. When you start your engine, avoid operation at high RPM until the oil pressure comes up to normal and the oil temperature rises into the green range on the gauge, which normally means that the oil has become viscous and is freely flowing throughout the entire engine.

We've all seen the pilot who allows the RPM to run excessively high right after the engine catches and begins to run as the starter is released. This not only wastes a small amount of fuel, but worse, that technique makes the engine run under high stress without the lubrication that it needs.

Remember that RPM means revolutions per minute, and revolutions without lubrication are damaging to the internal components of the powerplant. So the fewer revolutions that the engine turns during this crucial time, the less wear that is occurring. Put your hand on the throttle so that the instant you release the starter you can immediately ensure that idle won't be exceeded by more than 200 or 300 RPM.

After a few seconds operation near or at idle, you can safely bring the RPM up to about 1,000 to 1,200, which should be plenty to ensure that the alternator or generator is charging the electrical system; if not, have the electrical system checked by a mechanic. Allowing the RPM to "race" right after starting is ruining your engine faster than about anything else will, except maybe the equally poor practice of allowing the engine to suffer shock cooling.

Another side effect to improperly starting an engine in this manner is that the propeller might be damaged by sucking up debris from the parking spot, nicking the propeller blades, and increasing propeller maintenance costs. Most pilots never think about the propeller because it usually soldiers on for years without a whimper. Try to keep it that way. An overhaul of a constant-speed prop will set you back nearly $2,000, and overhauling a fixed-pitch propeller will still cost several hundred dollars.

If the outside air temperature is much below about 30°F, think about getting some preheat applied to the engine before starting it. Cold-weather starts are particularly damaging because cold oil is like jelly and can't lubricate worth a hoot until it warms up and can flow throughout the engine's tiny oil passages. The use of modern multigrade oils somewhat lessens the problems of cold-weather starting, but doesn't totally eliminate them.

Preheating the engine gets the oil warm enough to lubricate the moving parts much sooner than it could if left at ambient temperatures. There are two basic methods of preheating. The first method is the use of an external heat source, which might be either a combustion or electrical heater that has large-diameter hoses that are stuck into the cowling to spread warm air over the engine. The other method involves the installation of a series of electrical heating elements on the outside of the oil tank or crankcase bottom. The elements heat the oil when the system is plugged into a normal household electrical current. Each method has good and bad points.

The external heater generally does a better job of warming the top of the engine, especially the cylinders, than it does heating the bottom where the cold oil is. Unless the preheat session is prematurely terminated, the entire engine will gain some useful warmth and the damage from a cold start will be significantly eased. When the cylinders are warm, there isn't much chance of frosting over the spark plug electrodes during a cold-weather start.

The electrical preheating system that entails the installation of heating elements on the engine is somewhat costly and adds a minuscule amount of weight to the airplane because the coils are permanently attached. Then you've got to be in a place where you can plug into a typical electrical outlet in order to use the system. The heated element method does a better job of warming the oil than the external heat method but at the cost of a permanent installation and with the inconvenience of needing a source of current close by.

If you frequently fly into the colder areas of the country, you might consider the electrical element preheating method. If you are based in warmer climes and only venture into the colder states on occasion, maybe you're better off using a portable, external heater. But regardless of which way you go, get preheat when the weather is cold, and your engine will thank you with many more hours of trouble-free service.

## THE HIGH COST OF FUEL

Regardless of how you calculate the cost of flying a lightplane, fuel is a major component of that expense. If your plane flies more than 100 or so hours per year, the cost of fuel is probably the highest single part of the per-hour total cost. In the days of fuel prices below $1 per gallon, which are gone forever, most general aviation pilots would simply cruise along at a power setting that approximated 75 percent of rated power and not worry about much else that would be considered "fuel-burn management" except making sure that they didn't run out of gas.

Because avgas now costs around $2 per gallon in most of the country and in excess of that in many places, fuel conservation has taken on a new heightened sense of importance, unless you own a refinery.

A pilot can do two things that will affect the amount of fuel burned during cross-countries. The first is to select a cruise power setting that is an acceptable compromise between the competing goals of saving gas and still realizing a decent cruise speed out of the airplane. The second is to understand fuel mixture management and how it affects the rate of fuel consumption in the engine installed in a particular airplane. Let's look at mixture first.

## Fuel/air mixture

Gasoline won't burn without air, and we all know that an engine can't run on air alone. So the two are mixed together in the induction system of the engine to make a fuel/air combination capable of combustion in the cylinders. There is a wide spectrum of ratios of air-to-fuel that will burn in the engine, but for the best operation of an airplane piston powerplant, this range is narrowed considerably over what would, at its extremes, just support combustion.

The term *rich* is used to describe a mixture that contains more fuel, relatively, than what is needed to make the engine run. A *lean* mixture is one where the amount of fuel being combined with air is approaching the least that will burn in the engine without doing damage. Airplane engines obviously have to operate through a wide range of altitudes with denser air at low levels and thinner air at high levels, so the ratio of fuel-to-air, or the mixture, needs to be varied to meet the atmospheric density of the air that the engine is breathing at any given altitude within that airplane's operating spectrum.

Because automobile engines operate primarily in a very narrow range of elevations above sea level, they don't have the same requirement. Aircraft engines are usually equipped with a control to enable the pilot to easily and quickly vary the mixture right from the cockpit, while auto engines can have their mixture settings adjusted only by a mechanic in a shop.

When the mixture is rich, or heavily laden with fuel, the internal temperatures in the combustion chambers of the engine will be cooler than when the mixture is lean. Most modern airplane engines are designed to automatically enrichen the mixture when operated at full power. When this happens, the rich mixture helps to keep the internal temperatures down, aiding in the cooling process. Because takeoffs and climbs occur at relatively slow airspeeds, the normal method of cooling the engine by ramming large quantities of ambient air through the cowling is compromised at the very times when the engine is being asked to produce its maximum power, and therefore is generating the most heat. The rich mixture that is automatically set at full power solves that problem.

It is imperative that every pilot become intimately familiar with the pilot operating handbook for the airplane that he is flying. The manufacturer's stated recommendations and methods of mixture control must be followed. This chapter should be viewed as only general information. Lots of myths and rules-of-thumb are legend in aviation. Most of them have their roots in operating practices that were valid at some time, but many have been eclipsed with advances in powerplant design, modern materials, and the experience of flying these engines for more than 50 years.

One of the old "rules" that pilots learned was to leave the mixture control in the full-rich position at all times when the airplane was below 5,000 feet. Most modern airplane manuals have departed from that practice and permit leaning of the mixture at virtually any altitude, depending solely on power setting. Remember the problems associated with full power operation, and read your manual carefully. Unless you're doing a takeoff from a high-altitude airport, where a normally aspirated engine can't produce full power anyway, don't lean the mixture at a power setting above 75 percent.

Another old operating technique was to reduce power as soon as possible after takeoff and after obstacles were cleared. This practice doesn't do anything positive for the engine and in most applications produces a negative effect. Almost all modern lightplane engines are fully rated to produce 100 percent of power continuously for their entire operating lives. Quickly pulling the power back after takeoff robs the engine of the automatically adjusted rich mixture that it needs to keep cool while running hard at slow airspeeds.

Together with the fact that reduced power will lessen the rate of climb, and therefore only prolong the time that the engine is running hard with less ram air going through the cowling and cooling it, there just isn't anything to gain by quick power reductions after takeoff. Older engines made before World War II generally had limited amounts of time, often 3 to 5 minutes, when it was permissible to run them at full power. These limitations disappeared with almost all lightplane engines manufactured since the war, but you can see that old pilot habits die hard.

After your climb to cruising altitude is complete, start thinking about leaning the mixture. Establish yourself at cruise power, allow the airspeed to stabilize, then begin the leaning process. Many lightplanes have an *exhaust gas temperature gauge* (EGT), but just as many do not. Let's first assume that the airplane that you normally fly is not so equipped. There is one time-honored method for leaning the mixture without benefit of an EGT that still is useful today. Simply start slowly pulling the mixture control back; try to note two events while pulling back.

At one point during the leaning, assuming that your airplane has a fixed-pitch propeller, you should be able to detect a very slight increase in RPM. When this point is reached, you have leaned the mixture to the point that the engine is at a best-power mixture, which means that for that combination of altitude, airspeed, atmospheric conditions, and throttle setting, the mixture is now at the ideal setting for the engine to produce the maximum power; that is why the RPM increased. If you want speed at the expense of any further reduction in the rate of fuel consumption, you've got it here.

Assuming that your goal is to further reduce fuel consumption, keep pulling the mixture back until you sense that the engine is starting to run just a little rough. This point exceeds the physical limits of the fuel-to-air ratio needed to produce good running of the engine. Reverse the process and enrichen the mixture only that slight amount necessary to restore smooth running. You're now about as close as you can get to the mixture that will result in maximum fuel economy. Remember that any appreciable change in altitude, sometimes just a couple of hundred feet, requires readjusting the mixture. Even the slightest change in the throttle setting mandates readjusting the mixture.

If your airplane has an EGT, you can be much more precise in mixture control. Again keep in mind the fact that the temperature within the combustion chambers of the engine varies with the mixture; a rich mixture burns cooler than a lean one. Because we can't conveniently measure the temperature of the inferno inside the cylinders, the next best way to determine how hot the engine is running is to put a temperature probe in the exhaust manifold, just outside of the cylinder, and measure the temperature of the exhaust gases coming out. That's what an EGT does.

Remember when we were adjusting the mixture without an EGT? When you got the mixture too lean, the engine started running a little rough, and the engine would quit if you continued to lean. Well, once the mixture is so lean that things start getting rough, the temperature of the exhaust gases starts declining because the actual combustion within the cylinders is beginning to stop. You'll see that when using an EGT there is a point in the leaning process when the temperature stops getting hotter as you lean and reverses to getting cooler if you keep on leaning.

You'll learn that there is a *peak* in the EGT, and that peak is usually associated with maximum economy in most engines. If you have an EGT, you need to be especially cognizant of the instructions in your engine manual. The manufacturer will spell out several different temperatures for EGT, depending on whether you want maximum power or best economy. Variations of as little as 25°F either side of peak are important because some engines can only be operated slightly on the rich side of peak, while others can be run at peak or even 25° or so on the lean side.

Proper mixture control associated with reasonable power settings will do the most that can be done to conserve fuel. I learned this lesson very clearly one day in 1968 when I was ferrying a Cessna 180 from Ohio to Oregon. Being a young pilot, and one at that time who was not schooled in the constant use of checklists, I forgot to lean the mixture during the first leg of the flight to Missouri.

I noticed when I got to the first stop that the airplane was burning an inordinate amount of gas, probably 5 or more gallons per hour than it should. Then it dawned on me that I hadn't leaned at all during that flight. In those days, avgas was only about $.45 a gallon, and I didn't stretch the leg to the point that fuel remaining got critical, so it was a cheap experience. Since then I don't think that I've ever omitted leaning from any cross-country flight.

Another thing that will require readjustment of the mixture control is any use of carburetor heat. While you apply carb heat, the hotter air that is introduced into the induction system is by the laws of physics less dense that the ambient air. Hot air is always lighter and thinner than cooler air, all other factors being equal. So any application of carb heat will instantly and dramatically alter the mixture. You will need to further lean the mixture while carb heat is on in order to restore either the best power or best economy setting that you had established before using the carburetor heat. Don't forget to readjust the mixture after you remove the carb heat.

## RPM

If your airplane is equipped with a constant-speed propeller, you can save quite a bit of fuel by correctly managing the relationship between RPM and manifold pressure. When a constant-speed prop is run in the flat pitch, or high-RPM, mode, the engine is operating inefficiently. It's very much like driving your car in low gear. The engine is capable of pulling out from a stop, but the RPM increases rapidly as speed builds, and the engine is screaming if you try to travel at any reasonable speed.

Conversely, a car engine is being heavily taxed if it is asked to start out from a standstill in high gear. Too much load is put on it before it can build up the RPM needed to carry that burden. The same thing applies to airplanes. That's why we use the flat-pitch high-RPM settings for the power demands of takeoff and climb and "shift gears" into a coarser pitch, or lower RPM, for cruise flight in airplanes that have constant-speed props.

When you learn the relationship between RPM and fuel consumption, you'll see that the engine is burning more fuel at a higher RPM than it does at a lower RPM because the engine is rotating faster, and each rotation means that the cylinders are being fed fuel to power the fire. Take a look at the section of the airplane's manual that details operation of the powerplant. You'll see that there are usually several combinations of manifold pressure and RPM that will produce the same percentage of rated power.

If your manual is thorough enough to also show fuel-consumption rates at these various combinations, you'll probably be amazed to see that the combination of manifold pressure and RPM for 65 percent of power, which shows the highest manifold pressure with the lowest RPM, will also have the lowest fuel-consumption rate.

Another of the old myths from the pre-World War II days was that an engine should never be run "over square," or with a manifold pressure that exceeded the first two digits of the RPM setting. This old rule worked to prevent an engine from operating at 24" of manifold pressure and 2,200 RPM. Those who still adhere to this idea will always have the first two digits of the RPM higher than the manifold pressure, for instance, 22" and 2,300 RPM.

Almost all modern engines are not encumbered with such limits anymore. The advances in metallurgy, manufacturing techniques, and modern lubricants make today's engines considerably different than their ancestors. Look at your manual and see if your engine can be run over square. If the manufacturer OKs it, do it and you'll reduce fuel burns quite a bit. But remember that everything has to be accompanied by some common sense and run the engine "over square" only to the limits shown in the book. Exceeding those limits and running at some nonsensical combination like 28" and 1,700 RPM could trash an engine quickly.

Another savings is associated with using the lower permissible RPM settings. This relates to the fact that recording tachometers don't really record time, but they mechanically turn over the hour meter as the engine runs. A recording tach is accurate, in terms of recording true time, at only one RPM setting. Most Cessna tachometers are recording actual time when the engine is running at 2,566 RPM. Run the engine any slower than 2,566 and you're recording less actual time on the tach than you're flying.

The next time that you fly in an airplane equipped with an electrical hour meter, such as a Hobbs meter, notice the tachometer's hour reading and the Hobbs's reading at the beginning and the end of the flight. Upon making a comparison of the two, you'll likely see that the tach shows less time than the hour meter because the tach was not recording time at 2,566 RPM in cruise; the RPM varied from the instant that the engine started until you pulled the mixture back and the engine stopped.

There is a vital word of caution concerning the operation of any engine that has a constant-speed propeller; this relates to the technique involved in changing power settings. The never-to-be-forgotten rule is: *When reducing power, always lower the manifold pressure first, then the RPM. When increasing power, advance the RPM first, then increase the MP.* Even though your engine can probably be operated "over square" to some degree, don't risk exceeding the allowed limits. Extreme over-square settings can cause catastrophic engine failure in less than a minute.

## Speed/altitude

The next considerations that affect fuel consumption are speed and altitude. In general, speed is a product of power setting, but this is not the case during the descent phase of a flight. A portion of the speed of the aircraft when descending is a result of gravity, and that portion is free. From the viewpoint of fuel economy, it makes sense to throttle back during a descent and let gravity maintain speed. A couple of techniques are used by pilots during descents.

Many folks just leave the power setting where it was during cruise and get a little extra speed during the letdown. That works all right and does result in a little "free-speed" increase. Even more fuel savings will be realized if you reduce the power as you enter the descent and fine-tune the power setting to maintain cruising airspeed while coming down. You won't gain the time that you would if you left the power up, but the small loss of time is quite out of proportion to the gain in fuel economy realized when power is reduced during a descent.

If you keep track of fuel consumption rates, you'll soon realize that the extra few knots of airspeed that you get out of the airplane, whether in cruise or in descent, come at a high price in terms of fuel consumption. Remember that drag increases exponentially with an increase in airspeed, and power must overcome drag. That's why it takes so much more power to give you that last 5 or 10 knots of airspeed at the high end of the airplane's speed range.

In order to climb at any reasonable rate, you must increase power over that used for normal cruise. While it is true that most high-performance lightplanes can climb steadily while at cruise power, the rate of ascent is lethargic compared to what will result from the use of a more routine climb power setting.

There are several permutations on two basic climbing techniques. The first way to go up is to advance the power to that stated in the airplane's manual for normal climbs and then fly at the best rate of climb airspeed. Doing it this way will get you to altitude in the least time, which is useful if you want to get through a haze layer, out of the low-level turbulence, or for some other reason that dictates climbing quickly. But it's terribly wasteful of fuel.

The second way to go up is a cruise climb. Most modern manuals will suggest a cruise-climb airspeed and often suggest a power setting as well. Cruise climbing is nothing more than climbing at an airspeed that is between a slower-than-normal straight-and-level cruising speed, but faster than the best-rate-of-climb speed. This second method of climbing is preferable because it accomplishes a compromise between the excessive fuel burns for the little forward progress over the ground that occur when climbing at the best-rate-of-climb airspeed, while producing a better climb rate than one would see if you try to climb at cruise power settings.

Cruise climbing is usually done with a power setting that is somewhere higher than normal cruise power. This cruise climb power might be around 75 percent or a little more if you normally cruise at 65 percent or less. Usually the airspeed will settle about halfway between the best-rate-of-climb speed and the normal cruise speed that you're used to.

Unless you've got a real need to ascend quickly, transition to a cruise climb at about 1,500 AGL after takeoff. You'll see a reduced fuel burn rate, while at the same time you'll pick up a good deal of extra groundspeed as a result of the faster airspeed. Remember that a 20-knot increase in speed is much more important at the lower end of the airplane's speed range because it results in a higher percentage speed gain than does a similar raw increase at the higher end of the speed spectrum.

In cross-country flying, percentages of speed gain correspond to percentage savings in time, and therefore cost, that are about equal. Climbing at cruise-climb speeds also results in a reduced deck angle, which means that you will be able to see ahead with less obstruction from the cowling, thereby increasing your ability to avoid other aircraft.

Aside from any concerns with fuel economy alone, there are three basic speeds that you should know for your airplane because the day might come when you need to avail yourself of the advantages of each.

The first of these is the *maximum-cruise speed*, which in most aircraft is obtained at an altitude where full throttle results in 75 percent of rated power. You can run at 75 percent at lower altitudes, but the true airspeed will not be as fast as it will if you know the maximum altitude at which your engine can produce 75 percent. Because the air thins with altitude, the airplane produces less drag as you climb, and that is why this one altitude will result in the maximum cruise speed.

Being able to go as fast as possible usually carries a pretty stiff penalty in terms of fuel consumption. But there might be a time when you need to get somewhere as quickly as possible, such as when the weather at your destination might be deteriorating, or darkness is approaching (if you're not qualified or current for night flight, or if the airplane isn't so equipped), or some other real need for speed presents itself.

The second basic speed, *best-range speed*, is the one that will enable the airplane to fly as far as possible on its fuel capacity. In most lightplanes, this is approximately 40 percent faster than the best-angle-of-climb speed. Seldom is the best-range speed given in airplane operating manuals because the marketing folks are loath to publish useful speeds that are this slow. Best-range generally falls around 45 percent of rated power, and if you haven't tried flying with a power setting that is that low, it's a surprise the first time.

The ideal altitude to achieve best-range is like it was for best speed. Find the altitude where full throttle only produces 45 percent of power. At that high altitude, the drag on the airplane will be lowered because of the thin air. But the problem is that in order to get that high, you need oxygen to breathe and remain mentally alert. So the compromise is to go as high as you can without oxygen, say 10,000 to 12,000 feet (about 8,000 or so at night because of the marked decrease in night vision with increasing altitude), and set up 45 percent power; you'll get the maximum practical range out of your airplane.

Maximum range might be needed if you have to fly over bad weather or if your fuel state starts getting low. When you are flying over heavily wooded terrain, water, the desert, or other inhospitable territory, you might need to go as far as you can between fuel stops. Flying at slower airspeeds, within reason, always gets better fuel economy. Flying at 55 percent power will result in 10 percent to 15 percent less speed than you'll get at 75 percent power, but you'll save between 25 percent and 30 percent of the fuel that would be burned at 75 percent power.

You should also know your airplane's *best-endurance speed*, which is the third basic speed. This is the speed at which the engine is burning the absolute least amount of fuel per hour and still producing enough power to safely keep the airplane aloft. It has no relationship to range or efficient travel; it is only related to the most time that you can remain in the air. Best endurance is achieved when the airplane is flown somewhat slower than the best-rate-of-climb speed. The best-endurance speed is vital if you need to conserve fuel during ATC delays or if the weather at your destination is improving but is not yet good enough for your arrival. Use it when you just need to kill time without worrying about how far you will go in the meantime.

## WINDS

We've seen that fuel economy is best at the higher altitudes due to the reduction in drag inherent with thinner air aloft. But these advantages can be outweighed by the wind velocity and direction under certain conditions. To achieve the best economy, you've got to take the winds aloft into consideration. As a general rule, wind increases in velocity with altitude, especially during the colder seasons of the year.

If the wind velocity at your chosen altitude is less than 15 percent to 20 percent of your airplane's cruising airspeed, compensate with power adjustments. If flying into a headwind component, increase the power a little to fly faster and raise the ground speed over what it would be if you flew at 55 percent power. If you're lucky enough to have a tailwind of that magnitude, you'll save gas by throttling back and letting the wind keep your ground speed up where it would be if you were flying in still air.

The rules change when the wind components increase to the point that they exceed 20 percent of the cruising airspeed of your plane. Now it's better to vary altitude to either get away from a strong headwind or to take maximum advantage of a tailwind. The advantages of flying high quickly disappear if the headwind component at higher altitudes approaches 20 percent of the cruising speed. In that case, you're better off to descend where the wind speed is hopefully slower and suffer the increased fuel consumption inherent at the lower altitude.

But when you've got a tailwind that is increasing with altitude, get as high as you can, but use common sense. Fuel burn at climb power is excessive. If the total length of the trip is a short distance, you can waste more fuel climbing than you'll save with the tailwind. If your trip is longer than one hour, you will probably gain economy by climbing, depending upon how quickly your airplane climbs. For trips much shorter than an hour, forget climbing very high to save fuel because the results will be counterproductive; you'll burn more fuel climbing than you'll ever save cruising for a very short period of time at a higher altitude.

## ROUTE OF FLIGHT

It doesn't take the proverbial rocket scientist to figure out that you'll fly more efficiently in a straight line between two points than you will by zigzagging all over the sky. The newer electronic aids to navigation, such as loran and GPS, have become a real boon to economical flying, especially in VFR operations. The VOR airway system has inherent course changes that result in any flight using more than two VORs being required to change course as it proceeds from one station to the next.

If you fly very many cross-country trips, figure out the cost savings that you'll enjoy by being able to cut 10 percent to 15 percent off the distances of your flights by going on direct routes between your airports of departure and arrival. You'll see pretty quickly whether a loran or GPS unit will pay for itself in reduced fuel consumption and reduced time on the airplane, let alone all of the other benefits that come with the newer navigation devices. Considering the costs of fuel and maintenance, you'll be surprised how quickly you can pay for a loran or GPS by being able to fly the straight-line course between airports.

Also don't forget that you can fly in a straighter line if you dust the cobwebs off of your dead reckoning skills. If you're flying on a nice day and you don't need to detour from the straight-line course to avoid airspace where you can't or don't want to fly, why waste fuel and time by blindly following the jagged course that results from flying from VOR to VOR several times? Fly in a straight line, and use the VOR as a safety net if you get lost.

## AUTOGAS

When I learned to fly in the mid-1960s, we were taught that automobile gasoline went with aircraft engines about as well as cyanide does with the human body. Another old tale has fallen by the wayside. The Experimental Aircraft Association and another private concern or two did years of extensive testing to determine if autogas can be safely used in lower-compression aircraft engines. The findings were that these older engines designed to run on 80-octane avgas do just fine on unleaded or leaded auto fuel.

Because there isn't any leaded auto gas left anymore, unleaded is the only alternative. The use of autogas in these older engines has also helped reduce lead fouling of the spark plugs and the cylinders. The 100-octane avgas has many times more lead than did 80-octane. Running straight 100-octane gas in an engine designed for 80 octane requires aggressive leaning and more frequent plug cleaning or replacement.

The supposed answer to this problem was the advent of low-lead 100 octane avgas. But even low-lead 100 has about four times more lead than 80 octane, so the problems still persist. Plugs foul, and valves go much more quickly. Because autogas is for all practical purposes available only in the unleaded variety, using it in 80-octane airplane engines alleviates the hassles involved with lead fouling.

There are some cautions to the use of autogas, however. First, you must obtain the grade of fuel specified in the supplemental type certificate (STC) that permits its use, which is obtainable from the EAA or other sources. *Never use any auto fuel that contains any form of alcohol*; this stuff is still deadly to rubberized components of the fuel system, such as bladders, hoses, gaskets, and whatever else might be made of rubber. Gasohol is OK in cars but forbidden in airplanes.

Alcohol has a chemical affinity for water, and the water will form in solution in gasohol and not separate out like it will from straight gasoline. Because water in the fuel has been the cause of many an aviation accident, that is another important reason to avoid gasohol.

If you seek the advice of your mechanic, you'll find that any particular mechanic is probably either well set in favor of using autogas or dead set against it. It seems as though everyone has an opinion about autogas in airplanes, with a good deal of the opinions founded in anything but fact. Before you use auto fuel, get with your maintenance facility, and talk about it. If you can, seek the advice of an engine overhaul shop in which you have confidence. In the last analysis, you'll likely get mixed signals and will have to use your pilot prerogative and make up your own mind. If you decide that autogas is for you, *you must purchase an STC before you can legally use it*. Call your insurance agent and personally review the insurance policy to ensure coverage when the airplane is flown with autogas.

Most airports don't sell auto fuel because the aviation fuel companies often contractually discourage FBOs from selling it. If you've got a private field, you can obviously buy whatever fuel you want. Don't even think about carrying autogas in cans from the filling station to your airplane. If you don't blow up your car in the process, you'll more probably end up with water or some other contaminant in the fuel.

The difficulty in obtaining autogas has unquestionably curtailed its use in aircraft. The long-term effects of the increased use of alcohol in autogas in order to meet ever-increasingly stringent environmental regulations will also hinder the ability to find suitable auto fuel to use in airplanes.

But for now at least, the minimal cost of the STC (usually 50 cents per engine horsepower) can quickly be recouped by using autogas if you can conveniently do so at your airport and if an STC is available for your airplane.

## FUEL RESERVES

Two sections of the Federal Aviation Regulations dictate the minimum legal amount of fuel that must be carried in an airplane.

FAR 91.151 covers the requirement for VFR flights. You must have enough fuel onboard to get where you're intending to make the first point of landing (considering forecast winds and weather), assuming normal cruising speed, and then enough to fly after that for at least 30 minutes in daylight or 45 minutes at night.

For IFR flights, FAR 91.167 gets a little more complicated. First, it naturally requires that you have enough fuel to get to the first airport of intended landing. From there you must be able to get to your filed alternate airport, and thereafter fly for 45 minutes in an airplane or 30 minutes in a helicopter. All of this is considering forecast winds and weather and reported conditions if they differ from the forecasts.

That's it for what the FARs say. Few wise pilots plan cross-countries with that little margin against running out of gas. Because the focus of this book is about VFR cross-country flying, we will not concentrate on IFR reserves but will concentrate instead on some commonsense approaches to safely planning VFR fuel requirements.

There is a good argument that VFR-only pilots ought to carry even more surplus fuel than an IFR pilot because more can happen to make a VFR flight engage in lengthy detours around weather or engage in other contingencies that aren't of concern to an IFR flight. In large airplanes, where the weight of full fuel tanks can be considerable, the economies of operation often dictate against carrying large quantities of extra fuel, which is sometimes referred to as *tankering*. But in lightplanes, the difference associated with the increased costs of operation at the maximum fuel weight aren't more than pennies per cross-country flight compared with what would be gained by skimping on fuel supplies and thereby saving only a minuscule amount of operational weight of the airplane.

In my mind, one hour of fuel beyond the destination is the minimum fuel reserve that makes sense at any time. Knowing your airplane's best-range or best-endurance speeds can give you even more cushion, but always figure your reserves assuming normal cruising speed.

Very few single-engine lightplanes can carry people in all the seats plus baggage and full fuel. Something has got to give. Two things should never "give": adequate fuel and abiding by the gross weight and center of gravity limitations of the aircraft. If your cabin load is fixed, meaning that you can't leave passengers or baggage behind, the only variable in your loading calculation is fuel, unless center of gravity limitations demand that you leave some people or their luggage on the ground.

Don't let the promise of good weather, no matter how realistic that promise might seem, cause you to shave fuel reserves. Unexpected bad weather is only one of many problems that can be encountered in cross-country flying. There are two kinds of people who have never been lost in an airplane: nonpilots and untruthful pilots. If you take off without adequate fuel, getting lost for a short while increases the experience from a funny story that you tell at the airport some years later into a life-threatening emergency. There could easily be another problem, such as an accident at your destination airport that closes it for a while, necessitating a detour to an alternate.

Note that the FARs don't even require alternate airports to be considered for VFR operations. Again, the legal minimums don't necessarily comport with good judgment.

Regardless of how clear and blue the sky is, every cross-country should have as a part of preflight planning a planned alternate airport to go to if the primary destination doesn't work out for whatever reason.

For night cross-countries, common sense dictates that you should carry a greater fuel reserve than you deem to be the minimum during the day. Lots more can go wrong at night, such as deteriorating weather from fog formation and the obvious fact that it's much easier to get lost or otherwise disoriented at night. I would increase the minimum daytime reserve of 1 hour to 1½ hours at night for an experienced cross-country pilot flying in unquestionably good weather. The night minimum reserve would be increased as much as 2 hours for an inexperienced pilot or if there is the slightest chance that weather might create the need for a detour in the planned route of flight.

## SHOCK COOLING

Lightplane engines are wonders of modern engineering and manufacture. They are so reliable that most pilots trained since World War II don't think twice about the possibility of an engine failure. That's not so good because engines can and still do quit now and then, but there is no doubt that engine failures are rare today. These powerplants also endure tremendous abuse on occasion; just go to any flight school or rental operation if you doubt this statement and see the misery visited upon engines by the neophyte, the careless, and the ignorant.

Despite all of their stamina, airplane engines have some weak points. Because almost all aviation reciprocating powerplants are air-cooled, they have to operate over a wide spectrum of temperatures; they do so as long as they are properly handled and not subjected to excessive abuse. One area of abuse that is all too common and can be easily prevented is shock cooling.

Shock cooling is the term used to describe a situation when the engine is put to the task of enduring a rapid decrease in its operating temperature, and generally that affects the cylinder temperatures. When you're cruising along at altitude and need to start a letdown, there is a right and a wrong way to do it. If your panel includes a *cylinder-head temperature* (CHT) gauge, take a look at it while still in the cruise phase of flight. If you begin your descent by chopping the power, and if you keep the airspeed in the cruise range while coming down, you'll cause a rapid decrease in the CHT reading.

Even in the summer, the air at altitude is cool. The engine is running hot when it is producing cruise power. When you rapidly reduce power, all of a sudden you're asking the engine to cool perhaps as much as 200°F in just a few seconds. Any piece of metal shrinks when it cools and expands when heated. Metal will crack if heated or cooled too quickly. I found this out to my chagrin one winter when I fired up my wood-burning stove too quickly and the fire box cracked. Engine cylinders will do the same thing. In an extreme case, even the crankcase will crack if cooled suddenly.

Avoid shock cooling like the plague. Make it a practice to treat your engine gently, and it will repay you with thousands of hours of service. Abuse it, and you'll be replacing cylinders long before they should need to be overhauled in the course of ordinary operation.

Start your descents with a gradual reduction in power. If you need to descend faster, don't just chop the power more; do something else. Good planning should avoid the need for rapid descents. We all get a little behind the eight ball once in a while, but don't let that cause you to abuse your engine. Instrument pilots can work with ATC to avoid a clearance that requires a rapid descent.

If you've gotten close to the destination airport and are still too high, you can always fly a few 360° turns before entering the traffic pattern to kill time and allow you to descend without reducing the power too far. If you fly a retractable-gear airplane, the landing gear makes a wonderful speed brake. Extending it while descending will increase the rate of descent and reduce the airspeed at a given power setting.

Wing flaps are a good speed brake to get down faster and avoid the need to chop the power to the point that shock cooling might occur. If you're up high and need to get down more quickly than normal, slow the airplane to the flap-extension speed and lower the flaps to at least half deflection. Putting the flaps down increases the drag, so the power that you've got is now needed to keep flying, so you won't be as tempted to chop the throttle. Also, the reduction in airspeed that you probably had to make to get down into the flap operating range results in the descent's taking less time. Both are positive results in that situation.

Make it a practice to always keep some power on during descents, except for while in the traffic pattern where power-off approaches are not only acceptable, but good practice in smaller airplanes. If you've got a manifold-pressure gauge, try to keep at least 15" of power during descents. If you need more rate or angle of descent, and if you're more than 1,000 feet above your level-off altitude, lower the landing gear, slow down, and apply some flap extension. An alternative is making a 360° turn to eat up the altitude.

The key to avoiding shock cooling and the engine damage that can result is to plan ahead, descend at a reasonable rate, and keep some power applied until well into the traffic pattern. You've got to let the engine cool over several minutes, not seconds. If you're doing touch-and-goes at the airport, the problem isn't as severe because when you chop the power to fly the base and final approach legs of the pattern, your airspeed isn't that fast, and the cooling effect is retarded as a result.

Shock cooling is a product of a preexisting hot engine, rapid power reduction, relatively fast airspeed, and the resultant rapid decrease in the cylinder temperatures. Because you know what causes shock cooling, you can prevent it and the severe engine damage that might result.

# 8
# Flight management options

THE PRIMARY GOAL OF ANY CROSS-COUNTRY FLIGHT IS TO GET FROM ONE place to another. But another goal ought to be to do it as efficiently as possible, consistent with safety. There are several ways to accomplish that, and to do it requires a little knowledge of some basic facts and simple mathematics of cross-country flying.

Many pilots go about making a trip by casually walking around the airplane to make sure that nothing is broken, climbing in, firing up the engine, taking off, and climbing to some arbitrarily selected altitude that for whatever reason just seems to be comfortable. When these same pilots approach their destinations, they throttle back, descend, and land in the same unplanned manner. Flying without bothering to figure out what's going to make the flight more efficient wastes fuel (money), might well result in the flight taking longer, and is not precise.

The cockpit of the airplane is not the place to try to determine how to best fly the trip. That exercise should be a part of any sharp pilot's preflight planning, and then that flier only has to update the plans as necessary while en route to take into consideration any in-flight condition that wasn't accurately forecast. We'll take a look at some elementary mathematical formulas that can be used in any airplane to increase the precision and efficiency of cross-countries.

If you're thinking about getting an instrument rating, you'll need to know these techniques to pass the written exam, so why not learn them now? Even if you never intend to seek IFR privileges, flying precisely and with knowledge of how to make your airplane operate as efficiently as possible will always pay off in dividends of lower costs.

## EFFECT OF WIND ON A ROUND-ROBIN FLIGHT

Efficiency is composed of four elements: *time, speed, altitude to be gained or lost,* and *fuel burned.* Speed is a function of airspeed and the wind, which in combination produce a given groundspeed. The simplest illustration of efficiency concerns only a consideration of the effect of winds aloft on the selection of a cruise altitude.

Let's assume that we are going to fly a round-robin cross-country of 200 total nautical miles, or 100 miles in each direction, with the wind exactly parallel to our course. At a lower altitude, the wind will give us a direct tailwind of 20 knots on one leg, and we'll naturally have a headwind of 20 knots during the return flight. Further assume that the airplane cruises at 100 knots, which is a nice round number to make this example easier to understand.

If our flight were made in mythical no-wind conditions, the total trip would take exactly 2 hours, or 120 minutes. But when we fly in the wind conditions assumed for this example, the outbound trip will be flown at a groundspeed of 120 knots (100 TAS + 20 knots of wind). Whirling the wheel of an E-6B computer tells us that the outbound leg will take 50 minutes. The groundspeed on our fictional return leg will be down to 80 knots, so it will take 75 minutes to get home, for a total of 125 minutes for this simplistic trip. We have learned that on a round-robin flight any degree of wind hurts our efficiency unless we can vary the wind's effect by varying altitude from one leg of the flight to the next.

Now let's see what happens if we choose to make the same flight at a higher altitude where the wind is blowing at 35 knots against our tail on the outbound leg and against our nose on the return. Groundspeed outbound will now be up to 135 knots, so it'll take only 44 minutes to get to the destination. Coming back we'll have a whale of a headwind that will reduce our groundspeed to 65 knots. To go 100 miles back home, we'll spend 92 minutes aloft. The entire trip will take 136 minutes.

This simple example doesn't account for the time spent climbing to the higher altitude, nor for the fuel burned to get up higher. Nevertheless, it's easy to see that the trip took longer up where the wind is stronger. This leads us to conclude our first principle: *On a round-robin flight (from point A to point B and back to A), fly at an altitude where the wind is least strong.* Wind always slows a round-robin flight, regardless of its direction. The stronger it gets, the more it hurts.

Remember our example? A 20-knot wind resulted in a 125-minute round-trip, while a 35-knot wind stretched the round-trip flight out to 136 minutes. If there were no wind, a 200-mile trip in an airplane that cruises at 100 would take exactly 2 hours, or 120 minutes. Wind never aids the cause on a round-robin, and the rule holds even if you're flying in a crosswind situation.

The only way to bend this rule is to fly at a different altitude on each leg of the trip. In our example, we could go up to the higher altitude for the outbound portion of the round-robin and take advantage of the increased tailwind, then come home at the lower altitude where the headwind is less. But for short flights like this example, you have to go further in your planning and see if the time spent climbing, with the increased fuel burn and inherent slower speeds during a climb, will pay off in terms of the total time that it takes to make the entire flight.

## ALTITUDE, SPEED, AND FUEL BURN

Speed costs fuel because it takes ever-increasing amounts of power to overcome the higher drag produced by more speed. We also know that fuel economy increases with altitude as we lean the mixture as we fly higher. But by how much? And is it worth it? In order to answer these questions and see if the savings in fuel from flying at a slower cruising speed is worth the time spent, do this:

*Divide the distance to be flown by the speed in order to find the time aloft. Then multiply the result by the fuel flow per hour to arrive at the amount of fuel that will be consumed.*

Every cross-country pilot has got to have a good and thorough working knowledge of how to use the performance charts contained in the operating manual of the airplane flown. Without that ability, any real attempt to fly efficiently and economically will be futile. Here are the formulas to use, once you've gotten the numbers out of your airplane's book:

*Distance divided by speed equals hours. Hours times fuel flow equals fuel consumed.*

To calculate the savings between different available altitudes and power settings, fill in these numbers:

*Higher total time (or fuel consumption) minus the lower total time (or fuel consumed) equals a number that, when divided by the higher total time (or fuel consumption), equals a percentage of savings over the higher total time or fuel burn.*

We must reiterate that the following general rules of thumb must be confirmed from the performance charts in your own airplane. But for most lightplanes, going through the two formulas above will usually result in these conclusions:

- In slower airplanes, the slowest practical cruising speed appears to be the most economical, and flying at higher altitudes is a trifle less expensive than at lower ones.
- In high-performance airplanes, the most economical power settings seem to be between 53 percent and 56 percent of rated power, depending on altitude, with the highest altitude again producing the greatest savings.
- Higher altitudes pay off more in faster airplanes because the larger engines burn more fuel to start with, and even the same percentage savings of fuel between

fast and slow airplanes usually means that more actual gallons will be saved in the high-performance airplane.

These general rules do ignore one element of the cost of flying. Flying more slowly means that more time is put on the airplane and its components during a given flight. If you rent aircraft with fuel cost as a part of the rental price, you have to decide for yourself whether you want to be a fuel saver and pay more money in hourly rental charges for your trip. If you own your own airplane, you'll be putting a few more minutes per trip on your engine by flying efficiently, but the savings in fuel might well make up for it.

# HOW HIGH TO CLIMB

If we assume that the weather and other factors allow us to choose practically any cruising altitude that the airplane can reach, we have to decide how high to climb in light of the distance to be flown. This involves a little more math.

Lightplanes don't have the same lessened fuel burns at altitude as jets do. If you fly much on airliners, you've noticed that even for short flights they climb as high as possible. Sometimes an airline flight of only a few hundred miles consists of a climb followed almost immediately by a descent. That's because a jet, in percentages, saves far more fuel up high than does a piston engine. A jet really guzzles fuel down low and is only economical when flying up in the flight levels. There certainly is economy with increasing altitude in a lightplane, as we have already seen, but not nearly to the extent as in a turbine-powered aircraft.

As a pilot of a lightplane, you have to decide two things:

- How high to climb for a given trip distance.
- When the extra fuel consumed by climbing out paces the savings from a relatively short time spent at the cruising altitude.

To be precise, you need to calculate the time that you will spend climbing and figure the gas that will be used during the climb to altitude. Then figure the fuel burned during the cruise portion of your flight. To be really exact, determine the fuel burned during the descent when the power will even be less than at cruise. Add it all up and see how much gas got converted to exhaust fumes.

Go through this exercise for some hypothetical flights of various distances and with climbs to different altitudes. Of course, the numbers will be different depending on whether yours is a high-performance airplane or a more lethargic one. Because high-performance airplanes tend to have greater rates of climb, the time spent climbing to the same altitude will be less in a Beech Bonanza than in a Cessna 172.

Doing all of these calculations will lead you to conclude: *Climbing to excessively high altitudes for a short trip will not result in much, if any, savings in fuel. The low-performance airplane will suffer more, or save less, in such climbs than will the high-performance airplane.*

One additional factor that influences altitude selection is the effects of turbulence. Even in light turbulence of the variety encountered from convective activity on good-weather days, the airplane's cruise speed will suffer about 5 percent or so purely from the rough air. If you can climb above the low-level turbulence without having to go excessively high, you'll slightly increase cruising efficiency plus be far more comfortable.

In most parts of the East and Midwest, you can generally fly above the low-level bumps at around 7,500, sometimes lower and sometimes higher. If the cruising altitude that you think you want is only a 1,000 or 2,000 feet below the tops of the turbulence, think about going on up into smoother air, enjoy the ride more, and gain some extra speed.

## DESCENT

Coming down from altitude presents a tremendous opportunity to economize. Pilots generally descend in one of two ways: leave the power at cruise settings and come down at higher airspeeds; reduce power and come down at an airspeed close to cruising speed. Which is more efficient?

Remember that speed always costs fuel. Drag is produced by the airframe moving through the ocean of air, and drag increases exponentially with any increase in speed; therefore, the pilot who leaves the power at or near cruise settings and comes down faster might think that the airplane is being flown economically because that practice increases the groundspeed during the descent and thereby reduces time. Unfortunately, that is not the case.

Flying at a slower speed near cruise will enable you to really throttle back and save fuel during any descent. The higher you were cruising, the longer you have to descend, and the more gas will be saved by using the throttle-back method. If your airplane has a constant-speed propeller, remember our discussion in chapter 7. Fuel burn will lessen even more if you descend at the combination of RPM and manifold pressure that produces a given percentage of engine power with the slowest permissible RPM. So when descending, roll back the engine revs as much as you can, always staying within the book's limits.

Don't forget to enrich the mixture as you get into lower altitudes; it's easily omitted. Because you've reduced the power as you start down, the cruise mixture setting won't immediately show any indications of being too lean, and that might be the case for quite a while during the descent. But don't be fooled because if you need to restore power, whether it's for a temporary level-off or for any other reason, you don't want to be in a situation where you add power to an engine running at too lean a mixture to support the added power demand.

If you've got an EGT, monitor it carefully during the descent. If your airplane isn't so equipped, just push in the mixture a little at a time every 500 feet or so as you come down. Be certain that the mixture is full rich before you enter the traffic pattern at an airport, unless the elevation of that airport is quite high. If you need to add full power at any time to execute a go-around, you have to have the mixture in full to avoid engine damage, unless you're landing at a high-elevation field.

When descending, don't forget the possible need to use carburetor heat. Carb icing can sneak up on you unexpectedly, and a long, low power descent is an ideal time to encounter it. If your airplane has a carburetor temperature gauge, monitor it closely for any indication that the temperature is getting into the freezing range. Most lightplanes don't have a carburetor temperature gauge; therefore, watch for the other signs of ice, such as reductions in RPM or manifold pressure.

Because you're descending, manifold pressure won't be static unless you reduce the throttle setting about every thousand feet. In order to keep the power at the lower setting that you want during the descent, you'll have to keep reducing throttle to keep the manifold pressure where you want it for the descent. If you have a constant-speed prop, you won't be able to detect any RPM decline, and the manifold pressure is constantly changing, so you can't depend on that gauge to tell you if ice is forming in the carburetor.

Excellent options are to use carb heat once in a while during the descent or increase power for a moment or two now and then to see if the engine is icing up. In either case, remember to adjust the mixture as required for either the application of carb heat or increased power.

Other flight-related calculations that are more complex than these abound, but most are not practical for lightplane flying. If you have one of the newer electronic calculators specifically devoted to aviation and navigation functions, you can get as exotic as you want in precisely figuring every segment of a flight. If you're like I am and still use an aged E-6B "whiz wheel" that is older than dirt, you will probably limit yourself to the important parts of the trip where you'll be spending the most time. An electronic fuel-flow meter and totalizer makes these chores into child's play, once you learn how to use it.

The advantage of playing these numbers games with the performance figures applicable to the aircraft you fly is in knowing exactly what that airplane will do under varying conditions. Flying is an exercise in precision, whether it's tracking the centerline of a runway rather than drifting off to one side, or airspeed and attitude control during crucial phases of flight. Knowledge of the numbers and how they affect the efficiency and economy of flight is no less important to any pilot who wants to fly right and continually improve all skills.

# 9
# En route emergencies

EVERY PILOT SHOULD BE TRAINED AND EQUIPPED TO HANDLE THE emergencies that can reasonably be foreseen during cross-country flight. In this chapter, we'll talk about engine problems, electrical system failure, rapidly deteriorating weather, and getting lost. Additionally, the chapter reviews how a series of small problems can mushroom and grow from inconveniences into real problems if not solved as they occur.

Remember the airline training philosophy that if the pilot is trained enough in emergency procedures, they cease to be emergencies and are just additional procedures. Because the major United States flag airlines are just about the safest means of transportation available, that approach must have some merit.

## ENGINE FAILURE

Modern airplane engines are technical marvels of reliability. Unfortunately, they do quit and can develop other in-flight sicknesses that don't result in immediate failure but present the pilot with a situation that must be dealt with before things go south in a big way.

Without question, most engine failures aren't really caused by the engine, but by the pilot. These failures are usually fuel-management errors; the pilot either runs the airplane completely out of fuel or mismanages the fuel supply in a manner that the engine quits when there is still some gas available, probably in another tank. Engines don't run very well on contaminated gas either. Most fuel-management accidents can be avoided when the proper preflight is performed, the sumps are drained, and the pilot observes the grade and type of fuel added to the airplane during fuel stops.

Learn the fuel system of your airplane intimately, and know how to extract every usable ounce of gas in all of the tanks. Some airplanes, like the smaller Cessna singles, can burn fuel from both tanks at once; others, such as most Pipers, demand that the pilot switch from one tank to the other and have no position on the fuel selector that allows both tanks to feed the engine simultaneously. Renter pilots who fly different types of aircraft from time to time need to be sure that they know how the system works on any given airplane being flown and then remember those differences from one flight to the next.

Make sure that your engine has enough oil of the right type before beginning any flight, particularly a cross-country. Some operators will fly training airplanes on local flights with the minimum permitted oil supply. Nobody should ever start a cross-country under that condition. All engines are designed to consume oil, some more than others. Never skimp on oil, especially if you are a renter pilot and don't get the opportunity to become intimately familiar with the oil-consumption characteristics of each airplane that you fly. Oil is the lifeblood of an engine, so before you begin a cross-country, add the amount necessary to comply with manufacturer recommendations.

Let's assume, for the first instance, that you're a competent pilot who has done everything reasonable to ensure that you've got enough good and pure gas; you also know how to manage the fuel system in your airplane. Now we've eliminated the greatest cause of all engine failures, and you've checked the oil and know that there's plenty. Still, mechanical problems can arise that can't be prevented by a pilot, regardless of her or his competency.

## SUDDEN ENGINE FAILURE

Engine failures happen two ways: sudden and progressive. Your options in dealing with the two are different, so let's first examine the sudden, unexpected time when the engine packs it in for the day.

The first rule is: *Altitude equals safety*. When the engine quits, you're flying a glider, probably for the first time beyond training flights. Your ability to take the time to analyze what avenues are open depends upon the amount of time available. Time gliding is a direct product of the altitude at which the glide begins. Selection of cruise altitude involves many competing factors of wind, trip distance, airspace restrictions or types, and common sense.

If your flight takes you over the flatlands of the Midwest, a safe forced landing can probably be accomplished from a lower altitude than if you're over mountains, swamps, or forests, where it might take a longer glide to reach a suitable place to land. Keep that fact in mind when you plan your altitudes. If you're flying through a mountain pass when the engine gives it up, you have few options. The same goes for low flight over forests or other inhospitable terrain. I want at least 2,500 feet of air between me and the ground, more if possible.

The higher the airplane is when a sudden engine failure occurs, the longer a pilot has to sort things out, try to get the engine started again if possible, and to select a landing site. Most lightplanes have a glide ratio with the engine out from about 7-to-1 to maybe 9-to-1. Fixed-gear airplanes glide more poorly than retractables do, primarily caused by the extra drag of the gear hanging out all of the time, and drag is the enemy of the glider pilot. What these ratios mean is that for every foot of altitude lost during a glide, the airplane will travel forward about 7 to 9 feet.

Therefore, if you're a mile high when the engine packs it in, you can glide for about 7 to 9 miles, or viewed another way, you've got a circle that is about 14 to 18 miles in diameter in which to find a place to land. When your altitude is 7,500 feet at the time of engine failure, your glide can cover from just a little more than 10 miles in a fixed-gear plane to maybe as much as 13 miles in a retractable, which expands the circle's diameter to 20 miles on the low side and 26 miles on the high side. That's just another reason that altitude equals safety, all other considerations being the same.

There is another reason that flying high gives you more options if faced with sudden engine failure: Altitude equals time in the glide. Most lightplanes glide at a descent rate of about 500 feet per minute. If you're at 5,000 feet when the unexpected happens, you've got 10 minutes until touchdown to deal with the problem. At 7,500 feet, the time for head scratching goes up to 15 minutes.

This 5-minute difference is crucial because you need a few minutes, 5 ought to be plenty, to go through the engine-failure procedures of checking mags, switching tanks, getting the mixture to full rich, and whatever else your airplane's manual calls for in the circumstances. Then, when a forced landing is inevitable, you've got more time to select the landing site, plan a pattern, and execute the landing.

While in cruise flight, don't lose a sense of the surface winds. You can always listen to the ATIS from a nearby airport, monitor the tower of a controlled airport along your route of flight, and check weather and winds with an FSS every hour or so. There are other cues to watch for, such as smoke from power plants, waves on bodies of water, and even the blowing of trees if the wind is strong. Regardless of how you determine it, know the direction of the surface wind at all times, wherever you are, while you're flying.

If you are faced with sudden engine failure, the first thing to do is to try to find out why it quit. Change fuel tanks, pull the carburetor heat on, and do a magneto check. There is a slim chance that one of these corrections might solve the problem and restore power. But don't waste time denying the inevitable if the engine doesn't start because you're using up valuable time (altitude) before you start planning for the upcoming landing.

Know your airplane's speeds for minimum sink and best-glide distance. The airspeeds are usually close to the best-angle and best-rate-of-climb, but not always. If your owner's manual doesn't specify them, go out and experiment to determine them.

Best-rate-of-descent airspeed can be discovered from doing some glides in calm air and observing the rate of descent as displayed on the vertical speed indicator. Try gliding at 5-knot airspeed increments until you discover the speed that produces the slowest rate of descent. That speed will keep you in the air the longest, but it won't produce the farthest travel over the ground.

The airspeed that develops the most distance in a glide is harder to determine with casual experimentation. The best way and place to conduct this test is in the traffic pattern on a day with no wind. Fly a normal pattern, and when the airplane is abeam the end of the runway on downwind, reduce power to idle to begin a true glide. Make sure that you mentally mark the places where you turn to base leg and to final.

By making several such approaches, all beginning at the same downwind altitude and at the same distance from the runway, and reducing the airspeed each time from the fastest practical gliding speed to a speed just safely above stall, you can see what speed produces the farthest glide before you have to add power to make the runway for landing, or have to add flaps to avoid going around.

Make the glides without flaps, and fly each approach at different airspeeds in 5-knot increments. We all go out and do touch and goes once in a while, so make the best use of that practice by learning something new that might save you someday.

Naturally, if these glide speeds are given in the airplane manual, use the manufacturer's speeds because they were determined through flight testing that was performed more scientifically than these home-brewed experiments that we've described.

Back to a real emergency: Next, immediately turn downwind, especially if you are within 2,000 feet of the ground. I've found through more than 20 years of flight instructing that most pilots can make a successful forced landing more often if a standard traffic pattern is flown to the landing site. Very few people are able to make a successful straight-in approach with a dead engine. We're all more used to the visual cues in a rectangular approach, so stay with the norm as much as possible.

By turning downwind, you're now on the downwind leg for wherever you're going to land. Don't forget that the "runway" might be either on your right or left. Start looking for the place to land to the left, right, front, beneath, and sometimes behind the airplane. If the only suitable place to land is straight ahead, again depending on your altitude, you might have to do a straight-in; so don't deny yourself an acceptable field only because it's straight ahead.

Immediately after the engine has quit, reduce speed to the best-glide speed by raising the nose, and in the process, you'll get a little more altitude out of the "zoom" climb when you're trading off excess airspeed for altitude. Then you can glide the farthest.

If you've got a constant-speed prop, pull the prop control all the way back, to the low-RPM position. While almost no propeller on a single-engine airplane can be feathered, pulling the control back to low RPM will considerably lessen the drag from a windmilling prop by reducing the pitch of the propeller blades and the drag that pitch in the blades produces; the drag reduction will extend your glide even more.

Try it sometime for practice when you do a power-off glide, and you'll be amazed at the difference; it feels as if the airplane started coasting downhill. Just remember to put the prop control back in to the high-RPM position before advancing the throttle to recover from the practice glide.

The selection of the place to land deserves some time. Try to find a field that is as smooth as possible. In summer, larger pastures and wheat or oat fields make the best choice, all other things being equal. Try to avoid a plowed field, and if you have to use one that is freshly plowed, try to land parallel to the furrows unless the crosswind would be strong. Most importantly, avoid fields that would require the close final to be flown over wires or other obstacles.

Sometimes pasture fields are used for pasture because they aren't good for much else. They can be rocky or so uneven that they aren't suitable for farm machinery to work in them. If you do choose a pasture, be a little more watchful of the surface conditions than you think you might have to be.

If you have to choose between a good shorter field and a longer one that would involve a dangerous approach, pick the shorter one. Not many people get killed from running into a fence or ditch at the end of the landing roll when the airplane is moving slowly and decelerating. Wires are hard to see from any real height, so be especially vigilant for these killers.

One old rule of thumb upon which there is much disagreement is to never turn your back on the field you've selected. The veracity of that rule depends on altitude. If you're well above a pattern altitude by several hundred or 1,000 feet or more, one 360° turn might be in order to stay on the downwind if you need to lose altitude to land in a field that is relatively close to you.

This is the reason that you should always be aware of the elevation of ground over which you're flying. You can't perform a normal approach if you don't know how high you are above the ground. Another practice technique to use is to fly an approach to an uncontrolled, less busy airport, and put a cover over the altimeter. You can always have, and should be aware of, a good general idea of the elevation of the ground beneath you by looking at the chart. But you'll never know the precise elevation of the field into which you're making a real forced landing.

Practicing a few landings now and then without reference to the altimeter gives you a keener sense of judging altitude in the pattern and when to make the turns. It's a standard part of a glider pilot's training, but they expect to land away from airports and need to hone this skill pretty well.

Once you fly the rectangular approach and are certain that you have the field made, start adding flaps slowly. Don't add them early or all at once because you can never get back the altitude you'll lose at the steeper glide angle that flap application will produce. When you're practicing touch-and-go landings, try some approaches in which you have the engine at idle and use the flaps to modulate the angle of descent.

Flaps are one of the controls on the airplane, and every pilot should learn to use them to make the aircraft do what is desired. Application of flaps should never be by rote, but only as the product of a reasoned decision that their effect is needed. Glider pilots have only the glider's spoilers or speed brakes available to control the angle of descent, unless the aircraft is slipped. Yet few gliders undershoot or overshoot their intended touchdown points because their pilots learn quickly to use the available controls to get the job done.

Few pilots learn good slipping technique today because almost all lightplanes have effective flaps. In the days when airplanes didn't come equipped with flaps, slipping was the only means available to increase one's angle of descent beyond whatever angle resulted from the normal wings-level gliding speed.

Airplanes can still be slipped, and you should learn how to do it. Read the airplane's manual first because some aircraft, particularly Cessnas, are placarded against slipping with certain flap deflections because the extended flaps can blank the airflow over the tail, causing a tail stall if the airplane is slipped with flaps hanging out.

Also, know whether slipping is allowed in your airplane with certain fuel-load situations. Again, a hazard can result in some aircraft because the unlevel wings can result in a condition where the fuel pick-ups in the tanks become uncovered in the steeper banks that slips produce when the fuel supply is less than full, resulting in an interruption in fuel flow to the engine. Just read the book, then learn to slip in accordance with its permissible limits.

The idea is to contact the ground at the slowest possible speed, and flaps will help you do that. If you don't normally make full-stall landings, you should in most instances anyhow. Lightplanes land best when stalled, and being able to do a good full-stall landing is vital to a successful forced landing.

If you have to put the airplane down in trees or in a swamp, don't panic. Stall "land" a foot or two above the tree canopy or water, but no higher than that. You don't want a stall to develop that pitches the nose down. By stalling just into the trees, the forest canopy will attenuate most of the energy, and your survival chances will still be very high.

Make sure that seat belts and shoulder harnesses are worn and tight before the landing. Remember that the properly restrained human body can survive tremendous impact loading, even as high as 40 Gs for a short period of time. If you don't have shoulder harnesses in your airplane, get them. Nothing will do more to prevent injury and improve the crashworthiness of your plane, other than a good landing in the first place.

When you're on the downwind leg to your landing site, you've got to pay strict attention to flying a pattern that will get you to the planned touchdown point. The base leg, particularly the turn from downwind onto base, is crucial. A good rule of thumb: When the touchdown point is at a 45° angle behind your wing, it's time to turn base.

During the base leg, watch how the descent angle is going and see if you need to steepen it with a little flap application or slip, but don't overdo it. The last thing you want to do is to get too low on base and then be psychologically motivated to hurry the turn from base onto final, which is the time when most stall/spin accidents occur. Steep turns, low to the ground, are recipes for disaster and kill several pilots each year.

If you do fly the base leg erroneously and get too low, it's far better to accept the consequences of landing somewhere other than where you'd planned than to risk a stall/spin by steeply racking the turn around from base onto final. The normal inclination is to hasten the rate of turn by applying excessive rudder inputs.

To prevent the bank angle from getting too steep, the fatal scenario includes the use of opposite aileron pressure. A pilot who pulls the nose up to stretch the glide is in peril; a cross-control stall occurs in the bat of an eye, followed immediately by a spin. At the altitude of a normal base-to-final turn, nobody recovers from the spin.

One bit of education that I would recommend to every airplane pilot is to take at least a few dual flights in a glider. While you might not become addicted to soaring as a sport, even a small amount of indoctrination to glider flying will teach you more about energy management and planning than any number of printed pages or hours of discussion. Once you know how to routinely make a no-power landing, the fear of engine failure in power planes shrinks tremendously in your mind. Lack of fear makes any unusual procedure all the easier.

A far more critical problem rears its head if the sudden engine failure occurs during and shortly after takeoff. This is a situation where you've got to be able to react instinctively because there is virtually no time to think about what to do next.

An engine can completely fail, or start to fail, while still on the runway before the airplane has become airborne. This scenario is a no-brainer; chop the throttle, get on the brakes, and stop, even if it means accepting an overrun and going off the end of the runway. You'll be decelerating when you depart the runway, and because you haven't even reached liftoff speed before aborting the takeoff, you shouldn't be in much danger of a serious accident. If the engine starts running rough or coughing, or if it has any other indications of trouble, abandon the takeoff right now. Never try to achieve flight when the engine is telling you that it doesn't want to fly.

If the failure happens after taking off, your choices depend solely upon altitude. Another of the old adages of aviation is to never turn back to the runway, and that is generally good advice. Depending on how high you are, you can probably turn at least a little, and if a 20° heading change will take you to a better landing site than is available absolutely straight ahead, such a slight turn can be performed from just a few hundred feet up. Maybe you've got room to turn 45 degrees, or maybe even 90. It's purely a matter of judgment and experience. If in doubt, don't turn, and accept what lies ahead.

There is a point where a 180° turn can be safely accomplished, but that point is often when you're so high and far from the end of a relatively short runway that after the turn you won't make it back to the runway. But if the runway is long, like at a major airline airport, you've got a chance to get back if you're high enough before beginning the turn because you lifted off with lots of runway still ahead.

I've experimented with every airplane that I've owned to find out how much altitude was consumed in lifting off of the runway, climbing at full power, chopping the engine, getting the nose down to gliding attitude, doing a 180° turn, and then figuring that I needed at least 200 feet of additional altitude in which to get sorted out after the turn and do a landing. The results varied from 500 to 1,000 feet, depending on the airplane's gliding characteristics.

But, these weren't unexpected, hair-raising emergencies; they were planned events when I certainly knew what was coming. Raise any altitude that you might so determine by a factor of at least 50 percent to account for the reaction time that is necessary to comprehend the real thing, and start your response to it.

Think about this problem anytime that you're tempted to ask for or accept an intersection takeoff. It takes only a few moments to taxi to the end of any runway, unless you fly at Edwards Air Force Base. Those few minutes might save your airplane and skin if you lose an engine on takeoff, when you'll have the full length of the runway to stop on, or if you're high enough, to get back to.

If you are high enough to successfully turn around, remember that you'll probably be landing downwind. If there is any crosswind, make the turn toward the runway into that wind, so you stay as close to the runway as possible, rather than getting blown farther away from it. Lastly, if you have time, cut the mixture, magneto, and master switches to lessen the chances of fire if the landing is not successful.

## PROGRESSIVE ENGINE FAILURE

By using the term "progressive engine failure," we mean a situation where it is obvious that the engine is going to quit in the imminent future, but for now is still producing at least some power. This can be caused by a severe oil leak, being just about out of fuel, or a very rough running of the engine that seems to be getting worse. Sometimes things like cylinders or valve trains can crack or break, but the engine can still run to some degree.

It's obvious that a progressive engine failure is not as immediately as dire as is sudden failure, but it's still an emergency. Don't deny what is happening and try to press on to an airport if there is any doubt about whether you'll make it, unless the terrain is so inhospitable that the success of an off airport landing is very doubtful. It's almost always better to land somewhere with at least some power remaining to aid you in the approach than it will be to do a real-life dead-stick landing, even if you think it will be at a better site.

If the engine starts showing signs of coming failure, try to find out what is wrong. Go through the same checklist—change tanks, enrichen the mixture, check the mags—and try to restore proper engine function. If you've got oil all over the windshield, the answer is obvious, but in many of these progressive failures, you might be able to get things back to normal if you keep your wits about you.

The hard part about this scenario is deciding when to consider the engine's health so ill as to demand a landing now, off an airport. There is no rule for this one; only the judgment of the pilot can call this shot. When you've made the decision to land, treat the off-airport landing just like any other forced landing.

Sure, you've got some more time to find a good field, but don't fly the approach like you would with a healthy engine. Once you reduce power on final to idle, don't depend on being able to add any to stretch the final. In this situation, the engine often won't come back, and you've got to land wherever the airplane is pointed.

Lastly, remember that you can never stretch a glide. Human nature seems to be to pull back on the control wheel when you get too low. Doing that only increases the drag inherent with the resulting slower airspeed and increases the descent angle. There is always the danger of a stall, which is the worst of all possibilities next to a spin.

If your approach didn't work out and you find yourself short of the landing site, there is nothing you can do about it except a reduction in flap setting if you are at least a few hundred feet above the ground when you recognize the problem. Keep the airplane under control, and accept the landing that you've got. As long as you contact the ground with the wings level in a proper landing attitude and under control, your chances are very high that you'll walk away.

If the engine does resume normal running after you go through the checklist, land at the nearest airport. Back in the 1960s, one of my buddies owned a Cessna 140. He was planning the proverbial trip to Florida in a few days, and when he was flying locally just a short time before his planned cross-country, the engine quit. He got it running again, and forgot about the experience.

When he was on his way to Florida, the engine's symptoms developed again over the mountains of eastern Kentucky. In a few more minutes, things went very silent, and he crashed in those inhospitable hills. His passenger wasn't seriously injured, but we buried our friend after he had lived in the hospital for about a week.

## ELECTRICAL FAILURE

The failure of any airplane's electrical system should be no great emergency to a competent VFR pilot, even at night. The ammeter should always be a part of your routine instrument scan, and if it is, you'll notice that the alternator or generator has given it up while you still have enough battery power to last quite a while.

The key to managing an electrical failure is to recognize what has happened as early as you can. The ammeter or voltmeter is one instrument that very few pilots routinely observe and quite a few don't understand. Regardless of whether your airplane has an ammeter or a voltmeter, it should be an element of your scan every time that your eyes check out the instrument panel. If you own your own plane, think seriously about adding a low-voltage light, which is an annunciator light that comes on when the charging mechanism, whether it's an alternator or generator, fails and the battery starts discharging.

Your battery is nothing more than a storage vat that holds electricity. If it's fully charged at the onset of a charging system failure, all you need to do is to keep that vat from running dry before you're on the ground. But if the failure goes unnoticed for any appreciable period of time, the vat will be depleted very quickly before you start doing anything about the drain.

Years ago I had a Piper Aztec that a few other fellows also flew. One of them was a former military pilot who, I learned to my chagrin, had been taught to turn off the generator switches as a part of his shutdown checklist. One morning I started out on a cross-country after this chap had flown the airplane the day before. Everything was going fine until I noticed after a few minutes that the needles in the OBSs of both VORs were starting to drift aimlessly from side to side. About the same time, the audio of the communications radios' volume started to go very quiet.

The engines were running fine, and the weather was good. I couldn't figure out what was going on until it dawned on me to look at the ammeters. Both were dead against the stop, indicating full discharge. Then I had to think for a minute why both generators would fail at once. Trying everything that I could think of, I looked down at the generator switches that are located at the bottom of the power quadrant, near the floor. Both were turned off.

With a flick of two fingers, all was well again as everything suddenly came back to life, the VOR needles stood erect as a statue, and we were on our way again. The rest of the flight was uneventful. But that was the last time that I didn't include the generator switches in my prestart checklist. A subsequent visit with my friend cleared things up, and that was probably the last time that he turned them off.

This story is funny to remember now, but if the flight had been short and IFR when the failing radios could have become apparent only during an instrument approach, it might not have been so humorous. That experience taught me the importance of regularly scanning the ammeters from then on. Don't let something like this happen to you.

If a failure of an electrical-system charging device occurs, start shedding electrical loads at once. Turn off all avionics that you don't need, which might be all of them. If you know how to navigate by pilotage and dead reckoning, you don't need a radio at all. Save the battery power for the end of the flight when you might need to talk to ATC. You might need the battery power if things get bad and you get lost and need a navigation receiver to orient yourself.

Conserve whatever electricity remains in the "vat," because the battery is now your only source of power. Recall that you don't have to worry about the engine continuing to run because airplane ignition systems draw current from the magnetos, which have no connection whatever to the alternator, generator, or the battery. Regardless of how dark or quiet the cockpit might get, at least the engine will run. But if you ever lose your alternator in your car, it's a different story because a car engine derives its spark from the normal alternator-battery electrical system. A car won't run for long on battery power alone.

You should know the approximate current-consumption rates of the electrically powered equipment in your airplane so you know what to turn off. Things that have electric motors and high-power lights consume the most current. Don't turn on any landing or taxi lights unless you're on a very short final approach and are ready to land almost immediately afterward. A landing light turned on while on downwind or base leg could easily run the battery completely down before you flare for the landing. If you have one of the older ADF sets that includes a little electric motor in its antenna, turn it off.

If you're in doubt about how much current any item takes, look at the circuit breaker that protects it. There is a number on the breaker (or fuse in an older airplane) that indicates the maximum amperage that can go through that circuit before the breaker opens or the fuse blows. While the maximum amount of current that can be put through the circuit isn't the same as is being drawn from the equipment on the circuit, it at least gives you an idea about whether these electrical devices consume large amounts of juice, or whether their appetites for electricity are rather scant. The larger the number on the breaker or fuse, the more current is probably going through that circuit. Turn off the hungry items.

Also remember that transmitting over a radio takes many times more power than just receiving. Don't talk excessively because that will run down the battery all the more quickly. Turn off strobes and an old-fashion rotating beacon; the electric motor in the rotating beacon consumes lots of power. If you have an intercom that is wired into the airplane's electrical system, turn it off. Your goal is to get rid of every source of power consumption that you don't absolutely need.

If you fly a retractable-gear airplane, you'll need some power at the end of the flight to get the gear down. If you're at all in doubt as to whether it went down normally, use the emergency extension procedure. Some pilots recommend lowering the gear early after the electrical system fails, thinking that by doing it then, you're more likely to have enough power in the battery to get it down and locked safely.

Use your own judgment, but I think that I'd rather have the ability to talk and navigate as long as possible. If you lower the gear while still in cruise, you're going to reduce your airspeed as well, which only prolongs the time that you'll be draining current from the battery. The battery "knows" only the length of time until it goes dead; it doesn't care how fast the airspeed is.

Every retractable-gear system in a certified aircraft has an emergency extension method. If you fly a retract without intimately knowing how to use the emergency system to get the wheels out in the wind, you're waiting for an accident to happen. None of the emergency extension methods of which I am aware requires the use of electrical power. Even though the manual way of getting the gear down might require a lot of cranking or pumping, I'll leave the gear up until I need it for landing.

There isn't a whole lot that a pilot can do to keep the generator or alternator from failing, other than to overhaul them with the engine at routine times. These devices that produce the electrical power usually fail suddenly and without much warning. But you can do one thing to help you deal with an electrical-system failure when it does happen: keep a well-charged and fresh battery in your airplane. It's a real temptation to try to stretch a battery's life to the limit, but there's a risk involved.

That risk is that a weak battery on its last legs can't give you the reserve power that you might need in an emergency. Batteries aren't particularly cheap anymore, especially for an airplane with a 24-volt system. But even the $200 or so that one costs isn't that much when spread out over the life of a good battery. If you notice in winter that your cranking power is declining, or that the battery seems to run out of juice and is just barely starting the engine, don't wait to replace it.

The possibility of electrical failure is the main reason for carrying a flashlight and hand-held radio as a part of your equipment for every flight. Then it becomes more important to realize that neither do any good without good batteries. I use my birthday each April as the date to replace flashlight batteries (and home smoke-alarm batteries). Consider giving yourself a worthwhile birthday present: a back-up battery pack for the hand-held. Carry spare batteries for the flashlight and hand-held on every flight.

Some people replace batteries in the fall so they are sure to have good batteries for the upcoming winter flying season because winter generally involves more night flying. There's merit to that approach. The point is to pick a date that you can easily remember and replace batteries before you need a flashlight and can't even get 1-candle-power illumination out of it. If you use the flashlight much for other purposes, plan on battery replacement twice a year.

## RAPIDLY DETERIORATING WEATHER

A high percentage of all fatal general aviation accidents are the result of VFR pilots continuing flight into rapidly deteriorating weather. That fact is a real pity because continued flight into weather that the airplane or pilot can't handle is a preventable situation. If you fly long enough, you'll be on a cross-country when the weather ahead will go to pot, and you'll have to do something about it. The first thing to do is to recognize the situation early and get out of there before it becomes critical.

If you don't take good recurrent training, you ought to, especially if you fly cross-countries. The FAA has required biennial flight reviews for several years, but they should not be the sole element of a pilot's recurrent training program. All cross-country pilots who do not have instrument ratings should get some instrument time under the hood with an instructor at least once a year. Instrument-rated pilots must maintain currency to ensure competency when bad weather pops up during a VFR cross-country flight. The emergency instrument work that you had to do for your private pilot flight test is the minimum amount of proficiency that should be maintained by every pilot.

Instrument skills are rapidly lost if not practiced. If you haven't had some time under the hood in the last year or so, you probably won't be able to deal with the real thing and save the lives of your passengers or yourself if you stumble into IFR conditions. Pilots who have instrument ratings must remain current by logging at least 6 hours of instrument time every 6 months. If they fall out of currency, they can accomplish the 6 hours within the ensuing 6 months without an instrument competency ride. Past that 12-month period of time, instrument-rated pilots can't fly IFR until they take an instrument competency flight with an instrument instructor.

This should tell you that the FAA thinks that even those with the rating can't go more than a year without dual instruction to see how much their instrument-flying abilities have declined. If you don't have an IFR ticket and have only learned the bare rudiments of keeping an airplane right-side up by reference solely to instruments, how long do you really think you can maintain that skill? Not for long, that's for sure, unless it's practiced.

Two types of deteriorating weather can have sudden onsets: fog and thunderstorms. If you fly in a coastal area, be especially watchful for fog to set in during the evening and early morning hours. It can form and appear in a matter of minutes along a route that seemed totally benign to dangerous weather conditions. Fog is especially dangerous because it can persist all of the way to the ground. You can't fly under it

VFR; don't even try. If you see a fog bank ahead, the only successful course of action will be to turn around and avoid the area where the fog is forming.

The Midwest frequently gets large areas of generalized fog in the early morning hours, right after sunrise. These areas can cover hundreds of square miles. If you hear a prediction of fog during your preflight weather briefing, don't chance it. Wait until an hour or so after sunrise to take off, even if the weather looks clear. If the fog has formed, it will probably burn off well before noon, but burnoff rates can differ dramatically. Always check the weather thoroughly before assuming that your entire route has cleared of fog just because your departure airport is now enjoying VFR conditions.

The only good thing about fog is that the layer is usually only a few hundred feet thick. If you do get caught in fog at low altitude, use your instrument training to climb out of it. You should be back in clear air in a minute or so where you can at least maintain safe control of the airplane. You'll be stuck on top, but that's better than aimlessly wandering about in the clouds. Fog is also generally a localized condition, so if you're on top of a layer, a good old 180° turn, performed promptly, should take you back where you were before the encounter.

Thunderstorms are usually avoidable. Because you are VFR, you should be able to see a thunderstorm before you're dangerously close to it. You won't have the worry of encountering embedded thunderstorms that are brewing within a cloud mass. An IFR pilot flying in the soup without radar onboard the airplane might inadvertently penetrate a storm cell. When you do see a thunderstorm ahead, give it wide berth, at least 20 miles. The hail, high winds, intense rain, and turbulence can extend that far from the thunderhead cloud.

Never try to fly underneath a thunderstorm, even if the precip appears light and you can see through it to the other side. First of all, it probably isn't that light, and the intensity of the rain falling from a thunderstorm can increase in a few seconds. Secondly, again remember that the wind, turbulence, hail, and rain will be all around the thunderhead, and you can't see the effects of the wind and turbulence until you're in it. A microburst/wind shear is quite possible beneath the base of a thunderstorm. Attempting to fly underneath a thunderstorm is one of the unwisest things anyone can ever do.

If things start getting bad in a real hurry, whether it's from fog, thunderstorms, or general lowering of ceilings or visibilities, the only answer is to avoid the area. The time-honored 180° turn has saved countless lives in the history of aviation. Perform it early, and enjoy a safe trip to somewhere other than your planned destination. If you're caught in the weather to the extent that you can't get out by turning around, consider a precautionary off-airport landing. It's better than a crash from stumbling into IFR weather or having the airplane torn apart in a thunderstorm.

If you keep your wits about you and land in a decent field, you might do some cosmetic damage to the airplane such as breaking a wheel fairing, or maybe even dinging a cowling. That's all that's likely to happen. If you could accompany me to some of the wreckage inspections I've done in the past 20 years and see an airplane that now fits into the bed of a pickup truck as a result of someone who pressed on into the clouds, fog, rain, or a thunderstorm, you'd have a better understanding of the dangers involved in dicing with weather that you're not equipped or trained to handle.

## GETTING FOUND

Every pilot will eventually get lost. It's happened to me more times that I care to specify. The tricks are minimizing the occurrences and getting found after you're lost.

When I was 19 years old, I owned a Taylorcraft BC-12D that was one year older than I. Flying back from one spring break in Florida, my passenger was a buddy who was a commercial pilot; I had only a private license at the time. We were headed to Ohio and were crossing the Appalachian Mountains in the area of eastern Kentucky and southern West Virginia. If you haven't been in that part of the country, there isn't much there for checkpoints or visual routing. We got so lost that we didn't know what state we were in.

After wafting around for all too long, we spotted a freeway. Now we had a way to get found. We flew at a legal altitude and read the big green exit signs and finally came to one that said "CHARLESTON." We had enough sense to realize it probably didn't mean Charleston, South Carolina. We followed the exit road to Charleston, West Virginia, and landed for fuel. We only had to explain to the controller why we landed at a controlled airport without benefit of a radio. He quickly allowed as how that was better than the alternative of running out of gas over the mountains.

We were then on our way, only to run into deteriorating weather near Athens, Ohio. We landed at Ohio University Airport and took the bus back to Columbus. Quite a trip for a couple of college kids, but I'm here to write about it. The next time that I took that airplane through the Appalachians, I followed the West Virginia Turnpike without incident.

Modern electronic aids to navigation have significantly reduced the odds of getting lost, but it still happens. Radios and electrical systems fail, and so do lorans and GPS units. Keep track of your position over the ground all of the time; carry WAC or sectional charts for backup visual reference if you always fly IFR. Getting lost is most often a product of simple inattention.

Calculate your actual groundspeed as you progress along on the flight, and keep track of your checkpoints as you pass them. If you do get lost after knowing where you were recently, you probably aren't that far off course. Return to your planned heading, and keep alert for ground features that are recognizable. If you panic and start flying in circles or on divergent headings, things can go sour in a hurry. Using a navigation log will go a long way toward keeping you aware of your current position and progress during any cross-country flight.

Carry current sectional charts. Learn how to look at related ground features to positively identify a road, town, or other checkpoint as the one you think it is. Roads and highways go out of major cities like a spider web, and it's easy to pick the wrong highway unless it's cross-checked with something else. Also learn how to cross-check a position with only one VOR receiver; it's not hard at the speeds of typical lightplanes.

I've always found it easier to get lost in winter with snow cover on the ground. If you fly in winter climes, your navigation has to be much more precise, even over familiar territory. Areas just 20 miles from your home airport don't look the same with a blanket of snow. Pay attention, and you generally won't get lost.

Lastly, practice pilotage and dead reckoning. It's fun to fly 100 miles or more without using the radios to navigate, and it teaches habits that will come in handy someday. As long as you've got a chart, compass, and clock you should be able to go anywhere; if you can't, you're not a pilot anymore.

When reasonable time has passed, and you still haven't found yourself, don't be ashamed to get on the radio and ask for help, even if you have to use 121.5 MHz to get someone's attention. Just like my experience in Charleston all those years ago, it's better than the alternative.

# 10
# Climatic effects on cross-country operations

AIRPLANES FLY AND OTHERWISE FUNCTION WITHIN THE LOWER LEVELS of an atmosphere that has distinctively different characteristics as the seasons of the year change. It would be a good idea to take a look at some of the differences in the varying climatic conditions that will affect your cross-country operations and gain a working understanding of these effects.

## SUMMER

While most pilots tend to think of summer as the best season for all flying, there are some negative aspects to it as well. Granted, the days are long, airplane engines start more easily, and the velocity of the winds aloft is usually slower than in the colder months. But other characteristics of summer are less desirable for lightplane flying.

## Thunderstorms

Thunderstorms are the result of unstable air being lifted, generally quite fast, into the higher altitudes. The lifting force that spawns thunderstorms comes in two varieties: *convective* and *frontal*.

When the sun heats the Earth's surface, it does so unevenly because the varying features of the terrain absorb and reflect heat at different rates. Many people have the misconception that the atmosphere warms as the sun's rays pass through it on the way to Earth. The fact is that only about 10 percent of the total heat in the air results from the direct heating that occurs as solar energy passes through.

The other 90 percent of the total heat in the atmosphere comes from the energy that has struck the Earth's surface and then is reflected back up into the sea of air that surrounds us. The uneven heating of the surface, and therefore uneven reflection, result in areas of air that are warmer and cooler in relation to each other. When a column of warm air starts rising, convection is occurring, and a puffy little cumulus cloud will form at the level where the water vapor in that column of air cools and condenses into visible moisture.

Air cools with altitude because as it gets higher the air is farther from the surface where it's getting heated by reflection. There is always a point at which the water vapor in the rising air will condense, if there is enough humidity to condense. When an entire day is blessed with clear, blue skies, there isn't enough water vapor in the air to condense into a visible cloud.

If the air is unstable and has a considerable amount of water vapor in it, it's time to be on the lookout for thunderstorm development because the lower levels of the atmosphere will likely be unstable and because there is plenty of humidity under these circumstances to support considerable condensation.

That little cumulus cloud can grow into a thunderhead in a few hours with the classic anvil appearance at its top. The anvil is formed in the stratosphere where the winds aloft are always strong; those winds literally tear a portion of the top of the thunderhead away from the main body of the cloud, producing what we see from below to be in the shape of an anvil.

Convective thunderstorms often appear in the afternoon and early evening on summer days and generally form individually, rather than as part of a group of a line of thunderstorms. If the visibility is good in the general area, these storms can usually be circumnavigated without much difficulty. When you go around one, just remember that the rule of thumb is to avoid it by at least 20 miles in all directions.

The last chapter talked about how the violent winds, turbulence, rain, and hail that accompany thunderstorms can and often do extend well beyond the physical limits of the visible thunderhead. The danger of all thunderstorms lies in the violence produced within and around them. No small airplane can withstand the turbulence, rain, and hail.

In addition, even though your plane isn't electrically grounded, and therefore isn't an attractive target for a lightning strike, it's still possible to be struck while airborne if you're close to the thunderhead. Some lightning strikes result in only minimal damage, while others have destroyed the entire electrical and avionics systems in airplanes. There are a few unexplained accidents where investigators have theorized, but not proved, that the airplane exploded from a lightning strike in the vicinity of the fuel tanks.

If you venture directly into the storm, the chances are very good that your airplane will suffer structural failure, and you're out of options at that point. Even the Air Force recognizes this hazard and states that there is no peacetime mission that justifies thunderstorm penetration. If a fighter built like a bridge beam is at risk in a thunderstorm, think of your chances in a lightplane.

The more hazardous thunderstorms, if a distinction can be made, are those associated with frontal activity. In either warm or cold fronts, the warm air, which is often unstable, is forced upward over cold air beneath it to produce the lifting force needed for thunderstorm formation.

When a warm front is approaching, the system is often accompanied by widespread areas of low clouds and IFR weather. Thunderstorms that form within the general cloud mass and are referred to as *embedded* thunderstorms. The hazard here is that unless you have storm-detection equipment, you can't see them before it's too late.

Even though we're assuming that you're a VFR-only pilot, some pilots try to sneak underneath warm fronts if the weather is marginal VFR. Some of the time that tactic might have some merit, but if there is any forecast of embedded thunderstorms within the area of generally poor weather, watch out. You can be caught either in or below a thunderstorm before you know what has hit you because you can't see a thunderstorm embedded in a general cloud mass with anywhere near the same ease that a storm can be detected in otherwise clear skies.

The thunderstorms that come along with cold fronts are among the strongest and most violent. The reason is that cold fronts move much faster than warm fronts; therefore, as the cold air undercuts the warm air in front of the cold, the uplifting action occurs more quickly and with less opportunity for the different air masses to merge and temperatures to blend. The warm air is shoved up vertically much faster than in a warm frontal zone.

Lines of thunderstorms, called *squall lines*, frequently form ahead of cold fronts and race across the surface as fast as 50 knots. On rare occasions, squall lines can extend hundreds of miles in length, and lines between 50 and 100 miles are not uncommon.

The biggest danger that squall lines present to pilots is the fact that they can't be circumnavigated with any degree of ease or reliability. Even airliners with airborne radar usually detour as far as necessary to avoid squall lines. But they can do that because at their cruise speeds, going 150 miles or so out of the way means only a 20-minute detour. In a lightplane, such a maneuver means at least an hour and probably a fuel stop.

Another danger related to squall lines is the possibility of tornado formation. Tornadoes can form from almost any thunderstorm, but stronger and more violent storms are the best candidates for tornadoes. A tornado will almost always occur out of the rear part of a thunderstorm as the squall line or individual storm passes over a point on the ground.

The chances of surviving an encounter with a tornado are about as close to zero as anything can get. If a pilot is unwise enough to try to fly underneath the thunderstorm, just when he thinks that he's about out of the misery, he's entering the zone where a tornado could be lurking.

Don't play around with a squall line. If it's known to be in your path and close to your departure point, wait until it passes to take off. If a squall line develops while you're en route, land well ahead of it, and get the airplane either in a hangar or securely tied down without delay. The only nice thing about squall lines is that due to their speed of horizontal travel, they will usually move through an area quickly, and you can get back on your way to your destination without too many hours of delay.

## Haze

Haze is an annoying phenomenon, but generally isn't as dangerous as thunderstorms. Haze might mask thunderheads because of the reduced visibility, which is a real possibility, particularly in the eastern half of the country. Haze has increased since the 1950s in many areas of the East and Midwest, and its cousin, smog, is a well-known phenomenon for pilots in California.

The big problem that comes along with haze is poor forward visibility. Even though you can see straight down to a good degree, the slant visibility is awful, especially if flying into the sun. During your preflight conversations with the FSS weather briefer, if haze is mentioned or suspected, always ask to see if there are any pilot reports concerning tops of the haze layer. (Recall my previous recommendation to provide a pilot report about haze levels.)

If the sky is clear of clouds, it might be possible to climb through the haze layer and enjoy a flight in the clear, cool air above it. But there is a real danger in this course of action. Haze often tops out above 10,000 feet, where you can't fly without oxygen. That fact combines with the problem that if the reported surface visibilities are much less than 5 to 7 miles, you might be climbing through such thick murk that instrument flying is necessary to get on top.

If you have an instrument rating and obtain an airborne IFR clearance to climb, it's no big deal. But if you're restricted to VFR only, be careful if you start to climb and things get so mucky that you don't have a visual horizon. If that happens and you can still see the surface of the Earth, get back down to the point where you do have eyeball contact with the horizon, and accept the fact that you can't get on top of the haze. A one-eighty might be very smart if it can be safely executed.

Haze is even more dangerous at night, when your visual cues are fewer anyway. A VFR-only pilot shouldn't attempt night cross-country if the flight visibility is down to much less than 7 or 8 miles, and only then if you have plenty of options and have friendly, well-lighted terrain beneath you.

The loss of a visual horizon at anytime, especially at night, means that you are flying IFR. Spatial disorientation will quickly follow if you don't have instrument flying skills.

## Turbulence

The heat of the summer sun produces rising columns of air, and those columns result in *convective turbulence*. The worst of this thermally induced turbulence occurs in mountainous and desert terrain during summer afternoons. As much as glider pilots love thermals, power-plane pilots usually consider them a pain because they inevitably produce rough air.

In the flatlands east of the deserts, you might be able to climb out of the bumps at about 7,000 to 8,000 feet if the cloud cover permits. By all means, get into smooth air if this altitude is otherwise justified by your trip length and the winds aloft. True airspeed will increase some because the airplane isn't as efficient in turbulent air as it is in smooth air. You and the passengers will certainly enjoy the flight more than if the airplane were down in the low-level turbulence.

In the mountains and desert, you probably can't climb high enough in a lightplane to enjoy smooth air above the convective turbulence. Mountain and desert turbulence can easily extend up well above 10,000 feet, so you have only two choices. You can put up with it, which might get pretty tough, or you can do your flying early in the morning or early in the evening.

In both regions, the bumpiness can continue until quite late in the evening when you might not want to start a cross-country that will take you into darkness. Early mornings are much better, and the air won't be boiling until usually around 11 a.m. An early start gives you an opportunity to put several hundred miles under your belt before the turbulence develops.

## Heat inside the airplane

In this section, we're talking about the heat that builds up in the airplane when it sits on the ground, not the in-flight heat. Heat will reduce the useful lives of an airplane's interior, avionics, and exterior finish. On a sunny summer day, the interior temperature inside the cabin of an exposed airplane can easily reach twice the outside temperature.

Get a set of the commercially available window coverings that are made of "space blanket" or similar material, and carry them with you on cross-countries in the summer. Just because your airplane is hangared at your home base won't help a bit when you're hundreds of miles from home. Window covers work and only take a minute or two to hang in place.

If you don't have the full window set, at least get some kind of covering to put inside the windshield, above the instrument panel. Almost all glare shields are black, which means that they really soak up the heat from sunlight pouring through the windshield.

Radios are especially vulnerable to heat. Don't turn any of them on until a few minutes after the engine is started. The prop wash will blow some air through the cowling and avionics cooling vents to reduce the air temperature around the electronic components before they generate a certain amount of operational heat. Don't listen to ATIS before start-up because you'll only be further cooking the radio.

Here are some other ideas to further reduce cabin temperatures inside the parked airplane:

- Leave the cabin vents open. If there is any wind at all, it'll provide some air circulation, and even a little bit helps.

- Try to park the airplane with the nose into the prevailing wind.

- As soon as you get to the airport, open all of the cabin and storage compartment doors to let the cabin cool as much as possible before engine start.

- When ready to taxi, turn on only one communication transceiver and leave the other radios off until just before taking off or even until after you're airborne.

- Think about parking in the shade or in a hangar. Don't park underneath a tree unless you're ready to wash off all of the bird droppings that will result.

## Density altitude

While we're not going to go into great detail about density altitude and how it affects airplane performance, study your airplane's performance charts before flying in areas with hot temperatures and high elevations. Density altitude is not a measure of any true altitude, but it is an indicator of aircraft performance.

Even in central Ohio, where the elevation is around 800 to 900 feet MSL, density altitudes in the heat of summer are often above 3,000 feet. In the mountains and the desert, the density altitude can quickly get to the point that on a hot day the airplane's capabilities will be taxed to, and maybe above, the limit.

In you have never experienced really high density altitudes and the toll they take from a lightplane's performance, try a little experiment. On a normal summer day, find an airport somewhere in the Midwest or East where the elevation is no higher than about 1,500 feet MSL. Pick a field with a runway that is at least 4,000 feet long with no obstacles in the departure path.

Start to take off, but instead of applying full power as you normally would, take the power up to only about 2,200 RPM with a fixed-pitch propeller, or to around 22" of manifold pressure if you have a constant-speed prop. Stop your power application there, and see what happens. Naturally, be ready to apply full power if you need it.

You simulated the reduced power that will be available from the engine due to the debilitating effects of high density altitudes. The length of the takeoff roll will increase dramatically, and the climb rate after taking off will be lethargic. This trial doesn't really prove anything other than to show a pilot who has never flown in the "hot and high" what it might be like to take off with far less than the normally available takeoff power.

Don't try this experiment with any passengers or in any conditions where the slower lift-off and climb could be dangerous. After you see what happens at the lighter weights of the airplane with only the pilot as the sole occupant, imagine what it would be like if you had loaded the airplane to the gills with people, fuel, and baggage, and then expected the airplane to get off a short runway or clear any substantial obstacles shortly after taking off. Be thankful that your experience wasn't for real.

The density altitudes that appear in the summer are another reason to do any summer flying in the mountains or the desert in the early morning hours when the temperatures are still down to more acceptable ranges and the density altitudes haven't yet begun to soar. If there is any doubt about your airplane's ability to make any planned takeoff, be sure that you think it over carefully before attempting it.

If you decide to attempt the departure, pick a go-no-go point on the runway where you will absolutely abort the takeoff if you're not airborne by that point. That decision point better be a place where you will have ample room to stop before running off the runway.

Also be aware that the length of a landing roll also increases when the density altitude is higher because the thinner air produces less aerodynamic drag on the airframe; therefore, an airplane takes longer to decelerate. The increase in landing rolls is not as pronounced as the lengthened takeoff run, but this second effect of higher density altitudes is still present.

If you need a particular angle of climb to clear any obstacles after takeoff, study the airplane's performance charts carefully. Remember that the performance data was derived from the manufacturer's certification testing in a brand-new, well-rigged airplane with a new engine developing full power. An airplane with some time on the airframe probably does not have perfect rigging. The power output of an engine with more than a few hours on it has already begun the natural decline that comes with normal usage.

That factory test pilot probably had years of flight-test experience and knew what was coming each time a maneuver was performed. So under ideal conditions the airplane can match the book numbers, but you probably aren't faced with them. Add a fudge factor of at least 15 percent to the book performance numbers in any crucial situation in a lightplane. The fudge factor will compensate somewhat for the differences between you and your airplane and the test pilot and brand-new airplane that was flown for the certification flight.

## WINTER

The winter season of the year presents different problems to a pilot than does summer, but with adequate preparation and care, winter can offer some great flying conditions.

### Preflight planning

A winter preflight should be even more thorough than usual. This applies to the preflight planning of the cross-country trip and the physical inspection conducted on the airplane before starting out.

Winter weather changes more quickly than in summer because frontal systems literally race across the continent instead of proceeding at the snail's pace of the warmer times of year. Winds aloft and surface winds are stronger. Precipitation can take dangerous forms, such as sleet and freezing rain, that can turn an airplane into an ice cube in a terrifyingly short time.

Snow presents two problems. First, it reduces flight visibility by a factor many more times than does ordinary rain. VFR pilots should plan, as a general rule, to avoid all flight into snow because conditions probably won't stay VFR very long. The second snow problem concerns airports. You need to know whether the runways and parking areas have been cleared. Watch for snowbanks along runways and taxiways because they can ruin a wing or propeller that hits one.

You have a lengthening list of new concerns to investigate during your preflight weather briefing. Also ask about icing. No lightplane, especially one not equipped with ice protection and certified for flight into known icing, should even get close to icing conditions while in the hands of a pilot who is not instrument rated. An IFR pilot should get close only after *very careful* study of the entire weather system at play and only after plotting lots of escape routes in the event that icing is encountered.

Ice does lots of bad things to an airplane. It changes the shape of the airfoil on wings and propellers. When you're flying with any ice on the wings, you're an experimental test pilot testing an unknown airfoil. The propeller's ability to generate thrust decreases because it has also changed into an unknown shape. The ice sticking to the blades of the prop will soon cause it to go out of balance, adding additional stress to it and the engine.

Ice definitely increases the stall speeds, and the center of lift is different. Ice adds weight to the airplane. If you're already near gross, you'll quickly exceed it with any ice accumulation.

Lightplane windshield defrosters are anemic and can't deal with ice on the windshield; you'll be soon unable to see out the front at all. That enlightenment alone ought to be enough to encourage any pilot to stay out of ice. Table 10-1 presents the definitions for ice accumulation.

## Preflight inspection

Every flight should begin with a thorough preflight inspection of the airplane, but in winter, there are some additional things to check. Look into the wheel fairings for accumulated ice or snow; better yet, have your mechanic remove the fairings for the winter. The entire airframe *must* by regulation be clear of any frost, snow, or ice that might have fallen or formed on it. Snow probably won't be blown off when you take off, and frost won't melt off during subfreezing conditions.

If you put the airplane into a heated hangar to deice it, be sure that it is *completely dry* of melted frost, snow, or ice before you take it out into the cold. As soon as any melted water is exposed to subfreezing temperatures either on the ground or in flight, it will freeze solid as a rock, maybe even locking or breaking a control surface. To prevent this scenario, call ahead to your destination and reserve a heated hangar in which to store the airplane if you're going to remain someplace overnight.

Always carefully check the vents on the fuel caps and the static port to be sure both are free of obstructing ice, frost, or snow. Frozen fuel cap vents can cause an engine failure after takeoff, and a frozen static vent will cause your pressure instruments (altimeter, airspeed, and vertical speed) to be useless.

*Table 10-1. The definitions of icing.*

**Trace**

Icing becomes perceptible. The rate of accumulation is slightly greater than the rate of sublimation. It is not hazardous, even though deicing/anti-icing equipment is not utilized, unless encountered for an extended period of time, such as more than one hour.

**Light**

The rate of accumulation might create a problem if the flight is prolonged in this environment more than one hour. Occasional use of deicing/anti-icing equipment is needed.

**Moderate**

The rate of accumulation is such that even short encounters become potentially hazardous, and deicing/anti-icing must be used continuously or the flight diverted.

**Severe**

The rate of accumulation is such that deicing/anti-icing equipment fails to reduce or control the hazard, even when used continuously. Immediate diversion is necessary.

---

Thoroughly exercise all of the control surfaces, including trim tabs, to be sure that none of them have any frozen hinges. Water tends to collect in those hinges; if the collected water freezes, the ice essentially becomes a control lock that is not removed before flight, and you will be unable to control the airplane.

Engine starting is a particular nightmare in the winter. Because of weight considerations, airplane batteries don't have the reserve cranking capability that you're used to in a car system. You'll have to catch a good start quickly, or the battery will run down to a useless charge level before you expect it.

If the temperature is less than about 20°F, get some preheat. Engine oil naturally drains down off the rotating components of the engine and into the crankcase or oil tank after the engine is shut down. Oil congeals at colder temperatures; therefore, when the air temperature is ice cold, oil cannot restore adequate lubrication to vital parts in time for abnormal wear to be prevented. Preheating the engine warms the oil to more reasonable levels for proper lubrication when the engine is started.

If you do start a cold engine that fires a time or two and quits, you might have no choice but to preheat it. Water is a natural product of combustion. Those few short fires in the cylinders might produce enough water to actually frost over the gap in the spark plug electrodes, shorting them out. If that happens, further attempts at starting are futile; you've got to warm the engine and melt the ice that is blocking the spark.

## Taxi and takeoff

Before taxiing, you should carefully examine the surface of taxiways, run-up areas, and runways to see how much ice and snow are on them. If there is any wind of consequence, and there often is in the winter, watch out for slippery spots that could cause braking to be ineffective or unequal with one wheel grabbing on dry pavement and the other wheel sliding on an icy pavement. A sudden wind gust might push the airplane into a parked plane or hangar, and you might have no corrective braking control. Brake with caution unless you know for sure that the taxiway is totally clear.

Snow or ice on the run-up area might make it impossible to do a normal pretake-off check because the brakes won't hold the airplane when the power is applied. You might think about letting the engine warm up and doing the run-up while still tied down. (Make sure that the propwash won't damage anything behind you in the parking area.) Other alternatives include performing the magneto check by momentarily increasing power while taxiing in an area with good traction, or checking the mags early in the takeoff roll when the engine is at full power.

Taxi slowly. Be ready to encounter ice patches where the brakes won't hold. Be especially careful of snow piles along taxiways and elsewhere on the airport. Many airplanes, particularly low-wing airplanes, have come to grief when an inattentive pilot has hit a snow bank while taxiing. Snow banks freeze hard as rocks and will tear a wing or a propeller to shreds.

If there is any ice on the runway, be wary of crosswinds. You have very little aerodynamic control early in the takeoff roll and late in the landing rollout; you are depending on the traction of the tires to keep the airplane on the runway. If there is ice, there is no such traction, and you can be blown off the runway. Even after you reach the speed where the control surfaces become effective, you are doing a real dance to keep things in line. Read your manual, and know the proper placement of the controls during a crosswind taxi and/or takeoff. If at all in doubt about the runway condition, start the takeoff roll on the upwind side of the runway, rather than in the center, just to add a little more room if you start to slide.

## En route

Once aloft and on your way, the degree of extra vigilance required in winter cross-countries depends on the weather. If the visibility is good and the temperature is well below freezing, all the better. Still, listen to the hourly weather sequence reports because weather systems move very fast in the winter months, and forecasts are not as reliable as in the warmer seasons. The changes in the weather are also more pronounced, primarily due to the woefully reduced visibilities that come with any snowfall.

If any visible moisture appears, be ready to deal with airframe icing. Watch the wings' leading edges constantly, and if even a trace of ice appears, get out of there. You might be able to descend into warmer air and stop the ice accumulation, or you might have to turn around and go back to the ice-free air from which you just came. Don't push it. Ice will kill. If there is any freezing rain encountered, don't even think twice about continuing; do an immediate 180° turn, and make a hasty retreat.

As your flight progresses, keep abreast of conditions at airports that you pass. Your choice of an alternate or emergency field will be affected if any of them are closed due to snow accumulation. Many small and rural airports aren't well-stocked with snow removal equipment. Quite a few have to wait for the city or county road crews to finish the roads and then come out to plow the airport.

Never assume that any given airport is open if snow has fallen in the past day or so. Obviously, the same advice pertains to your intended destination, especially if it is a smaller airport. Always call ahead before departure to confirm the situation at all airports that you intend to land at when a winter storm has moved through the region.

When you're happily flying along over snow-covered terrain, it's very easy to get lost. I wouldn't recommend that any pilot try his or her first lengthy cross-country flight in the winter if he or she hasn't flown over white ground before. Many of the landmarks that make wonderful checkpoints in the summer can literally disappear in the winter.

Lakes and rivers are the most likely to become invisible if they are frozen over and covered by snow. Railroad tracks and secondary roads and highways can also get more difficult to spot. Small villages or towns that are normally discernible can fade into the snow-covered surface.

So, if your flight requires using any of these natural or man-made features for navigational checkpoints, be ready to search them out with increased difficulty, or be prepared to miss them altogether.

## Landing

The primary worry about landing is the runway condition. If you have any questions about it, perform a low flyby over the runway and have a good look before attempting to land. If there is any snow or ice on the runway, perform a soft-field approach and landing to give you the slowest possible touchdown speed to reduce the need for braking; remember that wind might prohibit or limit your use of soft-field technique. If the runway has clear approaches, plan your landing to touch down as close as possible to the threshold. That's good advice anyway because runways are made to land on, not fly over.

The danger of crosswinds is increased because the airplane can weathervane and you can lose control, especially as the airplane slows and aerodynamic control is lost. I would be very hesitant to try a landing in any appreciable crosswind onto a slippery runway. Try to touch down on the upwind side of the runway to give you more room for error (as also recommended for a crosswind takeoff).

If the runway is snow covered, you'd better avoid it. There is no way to even estimate the depth of the snow from the approach. If the snow is very deep, you'll likely break the nose gear, and a significant accident will result. Plus, snow depth can vary considerably along an open, wind-swept runway, and you could go from a skiff of snow to a drift more than a foot deep in an instant.

## Other winter advice

Cold temperatures will cause the air in the oleo struts and tires to lose pressure. During your preflight inspection, notice whether you need to have your mechanic add any air to either. While you can add air to tires yourself, most oleo struts require inflation by a strut pump that can generate pressures much higher than what are found in tires.

Also, many airplanes require nitrogen, rather than air, in oleo struts. Don't try to inflate a strut with an air bottle or tire pump; all that you'll likely accomplish is to lose the pressure that's still in the strut, and it will go completely flat. Then you're stuck for sure until you can get a mechanic with the proper equipment to do the job correctly.

If your airplane has been sitting for a while, you might notice a flat spot in the tire early in the taxi. Unless it's pronounced, the flat spot will go away as the tires become round again while you move along. If your flight has taken you from warm weather into cold, be careful on the first landing because the compressed struts and tires greatly reduce the shock-absorbing ability of the landing gear and also reduce prop clearance over the ground.

Check to be sure that you have the right weight oil in the engine. Unless your manual says otherwise, consider using the modern multigrade oils now available. Be aware that the oil pressure gauge needle will move slower after the engine starts and the oil eventually starts moving throughout the engine.

Inspect the condition of cabin heater hoses. They haven't probably been used for months and can develop cracks and leaks that could introduce carbon monoxide into the cockpit. Make sure that the cabin heat control works, too. In most lightplanes, it's a cable-pull device that can rust or corrode during the summer and not work when needed early in the winter; I've had that happen. One flight without heat will usually cause you to remember to check it before it is needed in the future.

The shorter daylight hours of winter make it far more likely that cross-country flights will either begin or end in the dark. Brush up on your night flying skills in the fall, and don't be caught needing to make your first night landing in months when you're at the end of a tiring flight. Make sure that all of your airplane's interior and exterior lights work, and get the flashlight out and put new batteries in it.

Wintertime is an excellent time to think about carrying extra survival gear. We're not implying that you have to load up for an Arctic expedition, but you should use some common sense. If you do have a forced landing, and all you have on are business clothes and a trench coat, you'll likely freeze.

Carry a parka and a blanket or two. Also carry boots or other footwear other than business shoes in case you have to take a little hike. Take a warm hat because most bodily heat loss occurs through the head. Pack some candy bars or other convenient food supply. If your route takes you over deserted territory, I'd carry a sleeping bag and maybe even a small tent and camp stove, if space permits. Preparations to this extent might seem silly, until you think about freezing to death.

Lastly, double your warm-weather fuel reserves. Winter reduces the number of usable alternate airports, and the weather changes much more quickly than in the summer. If you're flying into a headwind, realize that it will probably be stronger than you're used to in the summer. The greatest lifesaver of all might well be extra fuel in the tanks to enable you to deal with the unexpected.

Winter flying can be fun when you're knowledgeable and prepared. The airplane will perform at its best in the cold air. Visibilities can be astounding when the weather is good. A flight over the snow-covered plains or mountains can be as beautiful a sight as you've ever seen. Be prepared for winter and watch the weather carefully, and you will get quite a bit of cross-country utility from your airplane during the colder months of the year.

# 11
# Night cross-country

F LYING CROSS-COUNTRY AT NIGHT PRESENTS SPECIAL CONSIDERATIONS
that aren't present during daylight trips. Precise navigation takes on new importance, as does good planning, weather concerns, increased dangers of pilot incompe-
tence/complacency, and emergencies.

## ENGINE FAILURE

Every pilot who has flown at night has surely looked out of the window and wondered
what would happen if the engine quit. It's a very good question, but the numbers im-
ply that the wonderment isn't that big a worry. Because the modern aircraft engine is
such a reliable device if properly fed with fuel and maintained well, most of us accept
the slight incrementally increased risk of night flying in single-engine airplanes. I will
fly at night in a single-engine airplane, but I won't readily fly a piston-engine helicopter cross-country at night because the chances of survival in a night autorotation are
far less than in a forced landing in an airplane.

When considering the prospects of a forced landing at night, some pilots adhere to
the "dark spot" theory, which holds that a dark area on the ground is probably a field.
But it could also be a swamp, lake, woods, quarry, or other place where you'd rather

not go. There is no question that a forced landing at night is more likely to result in injury than one in the daylight.

But because life is full of risks and risk assessment, you'll just have to decide for yourself if the slightly increased chances of suffering harm at night due to engine failure are chances that you can take. There really is no good rule of thumb about where to head when the engine calls it quits at night. Perhaps a full moon would help you discern enough of the surface features to find a good place to put the airplane.

I don't believe that roads and freeways are the ideal place to put an airplane during the day. The hazards of dealing with overpasses, power lines and other wires that are strung over and beside roads, and the automobile and truck traffic have all strongly weighed against using most roads as an instant airport. Exceptions are highways through mountains, swamps, or forests, but for everyday planning, I've always assumed that a good farmer's field is preferable to a road. At night, that might be different.

The wires and overpasses are still threats, and they're almost invisible at night. But if I have absolutely no idea what kind of terrain or surface lies beside a good highway, I would probably opt for the pavement, hoping that I don't hit any wires on the way down. For the last few hundred feet of the glide, you could keep some extra airspeed to enable you to pull up and avoid an overpass if you see it in time.

Because the road is long and you aren't necessarily aiming for any particular touchdown spot, extra airspeed means some maneuvering ability that might just come in handy. Don't try to land on a median strip because they are full of steep gradients, drainage ditches, headlight reflector posts, sign posts, and other "gotchas" that you don't want to hit.

If you are landing on a road, land with the traffic, not opposed to it. Your gliding speed is almost surely faster than the speed of the cars and trucks, so you'll fly over the traffic as you come down. If the drivers are at all alert, they'll see you and slow down or get off the road.

Even your landing speed will probably enable you to touch down with very little speed differential between yours and theirs. When down and safely stopped, try to get the airplane off the road to lessen the chances of being rear-ended by an oncoming vehicle. Leave the strobes and position lights turned on as long as possible as an aid to being seen.

I limit my night cross-country flying in singles, but I maintain proficiency in twins. If the pilot of a twin is not on top of things, the statistics show that he or she won't do much better if an engine fails than will the pilot of a single who is now flying a glider. The pilot of a multiengine airplane might be more likely to have a serious or fatal accident.

In fact, of the accidents reported, twins that are flying on one engine are more apt to be involved in a life-threatening accident than are singles that have to do a forced landing. But these statistics don't reflect upon all of the times when a twin makes it to an airport on one engine without any further incident. The numbers don't tell the whole story because records do not reflect uneventful conclusions when the multiengine pilot does what he or she is supposed to do and lands the airplane safely with one engine operating and all other mechanical systems functioning properly. I'll take my chances in a twin any day.

Remember that the leading cause of engine failure, regardless of whether at night or in the day, is fuel mismanagement. If you have enough gas in the tanks and know how to get it to the engine, you've eliminated the biggest risk of power loss. Double all of your normal fuel reserves at night.

## GET SOME PRACTICE

Night flying is a rare event for the average general aviation pilot. I know of one who has 18,000 hours in his logbook, over more than 30 years, but only 500 hours of his total time are at night. That's not an unusual mix because most of everyone's flying occurs during the day, unless you fly for one of the night freight operators, or the scheduled airlines.

When you plan a night flight, and most of them will take place in the winter, get ready for it by boning up. The FAA's requirement of three takeoffs and landings every 90 days isn't enough to be anything other than barely legal, and "barely legal" is seldom equal to "safe." I've gone months without flying at night, and then when the occasion arises, I have to remember that I'm the same as a pilot who hasn't flown at all in months.

Don't be hesitant to get some dual instruction. The visual cues that we depend upon for takeoffs, approaches, and landings are greatly reduced at night, and some of them can play tricks on our senses. There are many times when visual references are so few that night flying demands some flight by reference to instruments, even if the weather is clear.

I've taken off from airports at night when the takeoff was into sparse terrain or over water, like Lake Erie. I encountered a total loss of the visual horizon until well into the initial climbout. If you can't fly instruments, beware of these situations. That's why the Caribbean nations don't even allow night-VFR flying; open ocean often doesn't present a useful visual horizon, regardless of how good the weather is.

Staying current for night flying can be difficult because with the advent of daylight savings time throughout most of the nation, few of us fly at night in the good-weather months of the year, which is when we tend to do most of our flying. I make a practice of renewing my night skills every autumn, as soon as darkness comes early enough that I can go out to the airport and fly for an hour or so without staying up into the wee hours of the morning. I will do that several evenings because one session is certainly not enough.

If you're instrument-rated, take along a safety pilot or instrument instructor because you will also probably have grown rusty in your IFR skills during the summer. Nighttime is perhaps the best time to do some instrument practice because the controllers are usually less busy, and you can work practice approaches, holds, and other instrument techniques into their flow of traffic much better at night.

# NAVIGATING AT NIGHT

Regardless of whether in the day, or at night, most cross-country navigation is done these days utilizing electronic aids to navigation. Night dead reckoning is the province of the practiced few and generally should not be attempted by most pilots. Even with a panel full of bells and whistles, finding your way in the dark has its own properties and peculiarities that differ from daylight navigation.

Visibilities are generally better at night, or at least they appear to be because the light from objects on the ground seems to be visible from farther away than the objects themselves are in the day. If you have any large antenna towers around where you live, you have probably noticed that you can see the lights on them from a long distance at night, and you probably cannot see the towers during the day from the same distance.

For that reason, you will be surprised during your first few cross-countries at night how close a distant checkpoint, like a town, appears to be. Don't be fooled; your airplane hasn't suddenly become that much faster. I've seen the general glow of the lights of a major city from nearly 100 miles away at night, especially at a high altitude. Distance can be deceiving in the dark, so be prepared for your eyes to play some tricks on you.

One of the best benefits about night flying is that you can see other aircraft much better from much farther away. This fact exists in large part due to the great improvements that have been made in aircraft lighting systems over the past few decades. I remember when a rotating beacon and standard position lights were all that most lightplanes had for anticollision lighting. In those days, it was quite a feat to have flashing navigation lights.

Thankfully things have improved to the point where most airplanes used routinely for night cross-countries have strobes, and some airplanes have external spot lights that illuminate the vertical fin. A good set of three-position strobes is about the most important and best insurance that you can buy for making yourself as visible as possible to other aircraft, and these lights aren't terrifically expensive.

Your selection of checkpoints will have to be different at night. You can't use road intersections, lakes, power transmission lines, and other such features that can't be discerned from a dark landscape. You're restricted to things that are lighted. Use towns, airports, major freeways, and the like as your checkpoints at night. The result is that there will be fewer places to see to check if you're on course; therefore, your attention to navigation needs to be all that more vigilant. Because it will be many times more difficult to "find yourself" at night, pay attention and don't get lost. If you're at all in doubt about your navigation skills, take an experienced pilot or instructor with you for a few night cross-countries until you gain the confidence you lack.

With the ever-declining price of loran and GPS units, anyone who flies very much night cross-country ought to invest in one, even if it is a hand-held portable. These gizmos can save a lot of sweaty palms, when you "aren't sure of your position."

During my college years back in the '60s, I ferried quite a few airplanes around the country. I once took one Bonanza from Columbus, Ohio, to Sioux City, Iowa, dropped it off, and flew another one back to Columbus, never seeing the light of day during the entire trip. You quickly learn about the peculiarities of night navigation earning your living that way.

## WEATHER AT NIGHT

When it comes to weather minimums, the Federal Aviation Regulations haven't been written to give enough deference to the differences between flying at night and in the daylight. If you take a look at FAR 91.155, you'll see that the minimum visibility is raised for night VFR flying in only two circumstances.

One is while in Class E airspace, when at or above 10,000 feet MSL, when you must have 5 statute miles of visibility instead of the 3 miles required during the day. The second is when in Class G airspace, higher than 1,200 feet AGL, and at or above 10,000 MSL. That's it for the regulatory increases.

In Class G airspace, you can still fly at night with less than 3 miles visibility as long as you've got at least 1 mile, if you're operating in an airport traffic pattern, clear of clouds. There are lower requirements for helicopters because forward speed can be slowed to a crawl, if necessary, in conditions of poor visibility.

The fact is plain that you just can't see and discern things as well at night. That's why the regs say you have to have 5 miles visibility in the two circumstances mentioned. Routinely flying VFR night cross-country in the minimum weather conditions allowed by the FARs is unwise. If the sky is overcast, regardless of how high the cloud deck is, you can't tell if it is lowering until you're in the clouds or so close that you might as well be in the soup.

Night cross-country under an overcast or broken layer is dicey stuff. Get a good weather briefing, and if the ceilings aren't high and stable, forget it. There is too much risk that you will inadvertently encounter IFR ceilings, and either be in the clouds, or forced to go too low, trying to stay out of them.

You won't know about the approaching danger like you will in daylight. There isn't any way to tell because the sky is black and featureless above you on a cloudy night. The bottom layer lurks above you, and you don't know where. An accidental penetration of the clouds by a VFR pilot is deadly anytime, but almost for sure will be in the dark.

Another thing that you can't see coming at night is an area of precipitation. At night, precip often means fog. If the moisture doesn't reduce visibility to the danger point, the fog will. If the clouds are low enough to be a concern at night, think about how well you'll fare if you blunder into rain and fog. When we talked about winter flying in chapter 10, we mentioned the debilitating effect that snow has on visibility. At night, it's even worse because you can't see it ahead in time to do something about it, even if that means turn around. If you get into a snow shower at night, you're IFR in seconds.

There are many times when a VFR pilot can fly during the day underneath a stable but lower ceiling. The same pilot can circumnavigate rain and snow showers in the light. You can't do either at night. The moral: Be extra conservative about night flying in any but the best weather unless you're instrument-rated, current, and flying an IFR-equipped airplane.

## ALTITUDE SELECTION

There is no time when the old saying that "altitude means safety" is more operative than at night. Flying low at night can be deadly, especially if carried to the extremes that you can't avoid obstacles such as TV antenna towers or high smokestacks. Until you've flown a fair number of hours at night away from the local airport environment, you just don't realize how hard some things are to make out; at the same time, other objects, like airborne traffic, are easier to spot.

Fly any night cross-country as high as is reasonable. The higher you are, the easier it is to navigate because reception of VOR stations increases with altitude. Communication ranges also increase with altitude because both are line-of-sight transmissions. If you have any problem that requires communications with ATC or an FSS, you'll find many more listening ears when you are high.

If you have any engine problems, altitude will be a real lifesaver. We talked in chapter 9 about how gliding distance increases with altitude, as does gliding time. When you have any sort of in-flight problem, it's compounded at night, and extra altitude provides you with all the extra time and options that you need to deal with the emergency. Because so few options are available at night to deal with an engine failure, adding even one more can mean the difference in the dark.

## HUMAN FACTORS

How you feel, how well rested you are, and your physical condition are factors that impact any flight, including a cross-country flight. These human factors that affect every pilot's performance take on a new and heightened importance during a night cross-country.

Because most people are active during the day and are not nocturnal by nature, watch for fatigue at night. Everything about night flying is more crucial, and if you're on your last legs at the end of a tiring day, you're not ready to conduct a safe night cross-country.

If the proposed night flight is a return to your home base from a day's business in another city, ask yourself how the day went. If you got beaten up during the day, do you really think that you ought to end a bad day with a night-flight home? Did you eat normally, or was it a day of missed or half meals? Is your metabolism up or down?

Everyone has dealt with "bad-work" days. I've flown to a court appearance or lengthy negotiation in another town and had the day go to pot. My morning meetings have dragged on until late afternoon. Something unforeseen occurs in the course of my one-day business schedule. I've dealt with an obnoxious or stubborn opponent in the courtroom. Dozens of things have happened to me that put me in a foul state of mind and depleted physical condition.

If that's how the day goes, I don't fly home that night. I've actually left a light-plane and come home on an airliner because I had a gut feeling that I shouldn't be flying myself. The price of an airline ticket or a hotel room is a small one to pay, considering the alternative.

It's difficult to overstate the debilitating effects that fatigue can have on human performance. Nobody thinks as clearly at night if he or she is worn out mentally or physically. Put that inescapable fact together with the certainty that night flying requires a higher state of mental awareness, and you should come to the conclusion that night cross-countries can be risky, even in a well-equipped airplane flying in beautiful weather, if the pilot isn't rested and alert.

Another problem with the human body is that we don't see as well at night. Our eyes are designed for daytime creatures, not night owls. Even though the FARs don't differentiate between night and day in the requirements for the use of supplemental oxygen, you should. Most pilots are fine for an hour or two at 10,000 to 12,000 feet in the day. At night, any altitude above about 8,000 feet should call for oxygen, and even lower than 8,000 if you are a smoker. Smoking reduces the normal oxygen concentration in the blood, and you need all of the oxygen you can get for good night vision. You've got a lot more to look for at night, and you can't do it with the naturally impaired vision that all people have in the dark.

We all start losing our near-vision acuity by or shortly after age 40. If you've reached that pivotal age, you'll have more trouble reading charts and other printed material in the dimmed cockpit-light environment at night. If you notice this, it's normal. See your eye doctor because you might simply need reading glasses for night flying a few years before you'll need them in the daylight.

Those folks with 20/20 vision who have always had good eyesight and who have passed their FAA physicals for years without glasses can often obtain the necessary aid from over-the-counter reading glasses. But just as I don't recommend doing your own medical diagnosis and treatment, I'd be careful about going this route.

Most of us don't drink alcohol and then fly, but a few pilots do. A look at the accident statistics shows that most drinkers who crash airplanes while impaired in the daytime are older, more experienced pilots. Perhaps they've been flying a long time, mixing it with drinking. To the contrary, those who kill themselves at night with alcohol in their systems tend to be younger and less experienced pilots. This might tell us something: You don't get away with any night flying if you've had a snort or two. The "daytimers" might get away with drinking before flying for years, but if drinking is tried at night, time would catch up with the pilot.

The FARs prohibit flying with a blood alcohol concentration at 0.04 percent or higher. Most state driving laws set the limit at 0.10 percent. Any alcohol consumption within the 12 hours preceding a flight should make you stay on the ground.

A major university did an FAA-sanctioned research project with a group of pilots (with safety pilots along) who flew during the day and night after being administered measured doses of beverage alcohol that was designed to induce certain blood concentrations. Some of the participants were given placebos, others received alcohol, and yet others were "staggered" (no pun intended) between the two extremes. Weeks of test flights showed that those who had only 1 ounce of alcohol at beverage strengths performed even the simplest of piloting tasks at dangerous levels.

## DELIGHTFUL DARKNESS

Night cross-country flying is a gorgeous experience. The lights that twinkle from the smaller towns and villages can look like Christmas 12 months a year. The atmosphere is generally more stable at night. The air is smoother because the turbulent effects of the sun heating the surface will most often settle down close to or just after sunset.

It is seldom very windy at night, unless there is a weather system approaching. ATC isn't as busy at most airports, so the level of radio chatter is greatly reduced. All of these conditions can combine into making a night flight about as peaceful and truly enthralling as any flight can be.

Just know which nighttime cross-country flying risks are different from daytime flying. Weigh them honestly, and use good judgment, and you'll be in for some real delights than can't be duplicated in the sunlight.

# 12
# The mountains

FLYING ACROSS TOWERING MOUNTAIN RANGES THAT REACH UP MORE THAN 2½ miles above sea level will be a daunting experience to pilots who learned to fly over and have only flown over the plains states, or anywhere else east of the Rocky Mountains. It won't be quite as daunting for pilots who have tackled the higher parts of the Appalachians, but even the Appalachians are no comparison to the truly rugged and expansive mountain ranges out West.

There is something awesome about mountain flying. The charts have that unusual brown shading all over them. The capabilities of the average general aviation pilot and airplane will be taxed, maybe to the limit. Some mountain operations will be beyond the capabilities of some lightplanes, and the pilot has to know when that dividing line between a taxing demand and an impossible demand will be reached.

You need to ask yourself about the service ceiling of your airplane and how long climbs will take as you approach the maximum altitudes at which your lightplane can fly. What are the commonsense rules for terrain clearance? Will you need oxygen? If you've never taken off from an airport as high as 6,000 or 7,000 feet, or higher, you're in for a surprise the first few times that you do it. Are there any special weather patterns or considerations about which you ought to be concerned?

Mountain flying requires expert chart reading skills and the ability to navigate without electronic aids. You will have to know the performance limits of your airplane and yourself. Unless you can do these things, you might well end up against the wall of a canyon, and maybe you'll be found in a few months by hikers or hunters, and maybe you never will be. Let's take it one item at a time and deal with each of these peculiarities of flying in and across the mountains.

# THE AIRPLANE

You absolutely have to be concerned with three performance parameters of the airplane you're flying: *range, service ceiling,* and *climb rates and angles.*

Flying in the mountains often means flying long distances between usable airports. Don't be fooled by thinking that private or public airports with short runways are suitable for use unless you're flying a Super Cub or other overpowered short-takeoff-and-landing airplane. If you're planning a flight in a modern airplane, you're going to need lots of pavement.

Design considerations in the development of general aviation airplanes have changed quite a bit since the 1950s. In those earlier years, lightplane engineers designed their wares for operations from short and unimproved airports. There weren't nearly as many paved runways available, and grass fields were considered the norm. Runway lengths were also a lot shorter back then, so airplanes like the Super Cub, Cessna 180/182, and similar older designs had much better short-field capabilities than do airplanes designed more recently.

Today's engineer of lightplanes thinks more about cruising speed, load-carrying limits, and creature comforts. Much less thought goes into short-field operations compared to the previous generation of aircraft designers.

The winds in the mountains are stronger than you normally encounter in the flatter parts of the country because you'll be flying at altitudes higher than normal. Winds aloft at higher-than-normal altitudes are generally of higher velocities; therefore, plan fuel reserves at least 100 percent greater than your normal practice to account for the slower ground speeds that you might encounter and the longer distances between refueling stops. Your leg lengths will naturally be shortened considerably, but fuel means the ability to make decisions and exercise options that aren't present if you're running on the fumes.

*Service ceiling* is defined as that altitude at which the airplane, at gross weight, will still climb at 100 feet per minute (fpm) at full power. That 100 fpm is a paltry climb rate that can quickly disappear if any turbulence or downdraft is encountered.

If you're lighter than full gross weight—you should be because you're always burning off fuel—you might get a slightly higher than book service ceiling, but don't count on it. Just like all of the other flight test data that were used to produce the manufacturer's book numbers, that data was gathered from flights in a brand-new airplane with a new engine, all in perfect condition. As the engine ages, it will put out less power.

Rate and angle of climb quickly become limiting factors in mountain flying. Most pilots never really consider the issue of angle of climb. Be sure that you know and appreciate the difference between rate of climb and angle of climb.

*Rate of climb* is the expression of how much altitude an airplane will gain in a given amount of time, usually per minute, irrespective of how far the airplane travels over the ground during that time. *Angle of climb* means how much altitude can be gained in a given distance over the ground, usually expressed in terms of 1 mile, not considering how much time is needed to travel that mile. When you need to clear an obstacle, it's the climb angle that becomes crucial, not the rate of climb.

Imagine taking off from an airport in the vicinity of Denver, the Mile-High City, headed west toward the "Rocky wall" of mountains that includes peaks up to and beyond 14,000 feet MSL, which is about 2.66 statute miles above sea level. You will need to know if the airplane will be able to gain 8,000 or 9,000 feet to clear the mountain range. If 360s, S-turns, or other detours are necessary to give you enough time to climb, you'd better know what is necessary in advance.

Before you venture into the mountains, know the best-angle-of-climb speed ($V_X$) as stated in the pilot operating handbook for the airplane to be flown. Because $V_X$ isn't used very often in normal flying, most pilots don't know it nearly as well as they do $V_Y$, the best-rate-of-climb speed.

Takeoffs from mountain airports can demand an angle of climb that some airplanes can't perform. Many airports are in valleys or are otherwise ringed by mountains, and the airplane has got to clear a granite obstacle in a given distance. There might not be room to maneuver, do 360s, or take other measures to extend the climbing time.

Be extra cautious in using any airport that appears doubtful. Remember that most airplanes that are based in mountainous areas are bought for their performance, regardless of other considerations. A Super Cub can get out of places that a modern airplane can't deal with.

Most of the tragic accidents that happen in mountain flying are the result of the pilot's overestimation of the airplane's ability to climb and clear the terrain. Be careful, and let caution be your guide.

## TAKEOFFS AND LANDINGS

You must experience a takeoff and a landing at a high-altitude airport on a hot summer day in order to believe the effects of less-dense air. No amount of reading and preparation will be an effective substitute for the real thing. The best way to learn what it's like is to make your first landing and takeoff from a large airport, like Albuquerque's Sunport, where the elevation is 5,352 feet MSL and the runways are long.

Because the runways are long at Sunport, you might not notice the higher TAS and groundspeed during takeoff, approach, and landing. What you will see, especially on a hot day, is that it takes forever to get the airplane to takeoff speed; then it will refuse to climb for an inordinate amount of time. When the climb does start, it'll be at only a fraction of the climb rate that a flatland pilot is used to. This kind of practice will instill a respect for mountain airports that no amount of chart and manual reading ever can.

It must be understood that while density altitude affects TAS and groundspeed, it does not affect the IAS and altimeter readings. No matter what the density altitude is, when the airport elevation is 6,000 feet, and you set the proper barometric pressure into the altimeter, it will read 6,000. If the outside air temperature is 90°F at that airport, the density altitude is approximately 9,200 feet.

A general rule of thumb is that a normally aspirated (nonturbocharged) airplane will require 25 percent more runway for takeoff and landing with each 1,000-foot increase in density altitude above sea level; thus, a Cessna Skylane that needs 1,350 feet to clear a 50-foot obstacle at sea level and 59°F will require a whopping 3,105 feet at a density altitude of 9,200 feet.

Again, don't forget that you and your airplane might not be able to duplicate the book performance numbers that were obtained during the manufacturer's certification testing that was performed with a new airplane and experienced test pilot. Add at least another 25 percent to what the manual predicts in any critical situation to compensate for the airplane and pilot differences.

Some neophyte pilots think that climbs and approach speeds should be increased in conditions of high density altitudes. Higher density altitudes affect the *true* airspeed, not the indicated airspeed. Rotate, climb, and approach at the same IAS as you would under the same loading and wind conditions anywhere else, but remember that the TAS and resulting ground speeds are much faster at higher density altitudes; therefore, liftoff and touchdown speeds will be at the same IAS, but much faster true airspeeds and groundspeeds.

Stalling speeds will be at the same IAS, but will occur at a higher TAS. For that reason, don't attempt a steep climb or steep turn shortly after takeoff. Keep the airplane level, and gain climbing speed slowly so that you aren't fooled by the visual sensation of the faster resulting groundspeed and think that you have sufficient IAS until you confirm it on the airspeed indicator.

If you need to leave a really high airport, particularly on a hot day, remember that you always have the option to take on only enough fuel to fly (with a reasonable reserve) to the nearest larger airport, or one at a lower elevation. In an extreme case, you'd better ferry the passengers out one at a time.

The airport at Leadville, Colorado, which sits at an elevation of 9,927 MSL, might be an extreme case for you. On a 90°F day, the density altitude will approach 13,000 feet. Because Leadville's runway is only 4,800 feet long, you most likely won't make it in most airplanes loaded anywhere near gross weight. Some aircraft don't even have a service ceiling that would produce *any* climb there under those conditions. If you have to land at or take off from an airport like this in the summertime, you'd better plan it for early morning when the air is cool and smooth.

Mixture control is vital to high-altitude takeoffs. So many of us are used to always using full rich when taking off, and it never occurs to us to lean the mixture this time because we're leaving a high-altitude airport. The engine can't produce anywhere near 100 percent, or even 75 percent power, in these conditions; so, after the usual runup, hold the brakes, apply maximum power, and then lean the mixture to the best power value on the EGT. If you don't have an EGT, lean until the engine begins to run rough, then enrichen it just enough to restore smooth operation. If you have a fixed-pitch prop, try to lean for maximum RPM.

The engine can't produce its maximum available power at high altitudes without leaning. In a tight situation, proper leaning before takeoff might just make the difference between success and an accident. Even if the case isn't that extreme, you run the risk of fouling the spark plugs by running at full rich at a high density altitude.

## INSTRUMENTS

In addition to the usual navcom equipment, a number of other instruments and electronic aids to navigation take on increased importance in mountain flying. In the high country, VORs are often few and far between, and the mountains between them can prevent reception; therefore, an ADF can be very helpful, if you know how to use it. In good VFR weather, standard commercial broadcasting stations can be a source of navigational information with the ADF.

An EGT is even more important to the mountain flier. While having one will result in more economical flying anywhere, you've got to be able to accurately lean the mixture for mountain operations. Even during the descent and approach to landing, proper leaning is important at these altitudes and will better enable you to get maximum power out of the engine in the case of a go-around, which is when you might need every ounce of available power that you can get.

One instrument that is not often found on lightplanes is an *angle of attack* (AOA) indicator. The AOA system consists of a probe, similar to the stall-warning probe that your airplane already has, which feeds data to a readout in the cockpit; the data are displayed by means of a needle that moves to indicate the wing's angle of attack in all situations. Mountain flying involves many misleading visual cues, including the cues related to the faster groundspeeds that result in the same IAS (detailed a few paragraphs ago).

Because the terrain is generally sloping in the mountains, there are many times when you won't have a level horizon for visual reference. You might think that your climb angle or other nose attitude is much different than it actually is because the true horizon is at the base of the mountains where you're not used to looking. If you encounter this sloping terrain, you can't depend on the aircraft attitudes that you normally use to produce certain performance parameters, or even to avoid a stall in some circumstances. Use the AOA or airspeed indicator.

When you get used to flying with an AOA, you'll wonder how you ever got along without it. The system is not particularly expensive, so ask your mechanic about installing one before you venture out West. Even in your normal flying, an AOA will tremendously increase your ability to get the best performance out of your airplane in takeoffs, climbs, and especially in short-field operations.

Get a GPS receiver or a loran unit. VOR signals are often blocked at the altitudes where lightplanes fly past and between mountain ranges. VOR isn't that reliable in the mountains unless you're up high above the terrain; unfortunately most lightplanes can't fly that high.

GPS and loran are not affected by line-of-sight restrictions. GPS is the wave of the future; therefore, if you're buying a new unit, it's probably wiser to invest in a GPS receiver instead of loran. Many excellent hand-held GPS receivers cost well under $1,000 (1994 prices); panel-mounted units are still fairly costly.

# OXYGEN

Oxygen will be needed in the mountains. Don't even think about serious mountain flying without at least a portable oxygen system in the airplane. Regardless of the legalities, prolonged flight at higher altitudes will result in a headache and a loss of mental alertness.

Flying in the mountains requires a degree of mental awareness much greater than the common flight in the low country. Things happen that are unexpected, and the pilot has got to be able to think fast. If you're groggy from lack of oxygen, you're playing Russian roulette with a revolver that has a round in every chamber.

If you can plan a route to stay below 12,000 feet, there is no guarantee that weather or some other problem won't force you higher. Clouds (that are obscuring ridges), wind, turbulence, or other factors could force you up to 15,000 or 16,000 feet, and without oxygen, you're sunk. Planning such a flight, without realizing that you need oxygen on board to deal with these contingencies, is haphazard, might cause you to fly at illegal altitudes, and might prematurely end the flight.

Portable oxygen systems that strap the tanks to the backs of the pilot and copilot seats are readily available from most of the aviation catalogue supply houses. Nasal cannulas that are similar to the type used in hospitals are approved for use below 18,000 feet. Use the cannula, and you won't have to put up with the discomfort and hassle of an oxygen mask at the altitudes you're likely to fly.

Very few normally aspirated airplanes can climb above 18,000 feet anyway, so discard any notions about the inconvenience of using a mask as the excuse for not having oxygen on board. These portable systems are not very expensive and make a handy optional piece of equipment to have for the times that you might want to fly at the higher altitudes, regardless of the terrain's elevation. They are especially useful for night flying when any pilot ought to use supplemental oxygen above 8,000 feet.

One word of caution is in order about the oxygen to use. Refill your system with aviator's breathing oxygen, not medical oxygen. Many health care professionals who fly lightplanes don't realize that they shouldn't take their airplane oxygen tanks to the hospital for refilling. Medical and aviation oxygen are not identical. Medical oxygen is formulated with a much higher water-vapor content to make it more comfortable for long-term use by persons who are on oxygen for days or weeks at a time, perhaps always on oxygen. The increased humidity of medical oxygen lessens the dryness in throats and nasal passages that a patient can suffer over the long-term.

You don't want the excess water vapor in aviation oxygen because the moisture takes up volume in the tanks, which reduces the time that the supply will last. If you're flying in cold weather, the humidity in medical oxygen can cause freezing in the lines and regulator of your airborne oxygen system when the airplane is sitting on the ramp in freezing temperatures. Refill your airplane's oxygen system at the airport, where aviator's breathing oxygen is available.

## SIGHT-SEEING

The very beauty of the mountain country can be a lurking trap for the unwary. Dropping down to altitudes with minimal ground clearance to see the sights is a very bad idea for any pilot who does not have an intimate knowledge of the terrain and is not flying an airplane with the climb performance to get out of a tight situation.

It might be exciting to see the cliffs and canyons from close-up, but many an airplane that is suddenly faced with rapidly rising terrain might be unable to climb fast enough to clear whatever is ahead and might not be in a position to do a 180° turn to get out of harm's way. If you want to see the sights close up, land at a suitable airport and take a ground-based tour. Consider paying the price to take a local operator's sight-seeing tour in the kind of airplane that has the STOL performance necessary to do it safely. An excellent compromise might be contacting a local flight school in advance and arranging for dual instruction in high-altitude flying combined with sight-seeing.

## WEATHER

The weather in the Rocky Mountains differs from that in all of the other parts of the country. Visibilities are usually astounding, ranging from 50 to 100 miles. Haze and fog are rarities, and the air pollution that plagues areas of both coasts is absent. The result is that a pilot who isn't used to being able to see so far has to relearn the ability to judge distances.

Sometimes a major mountain or other prominent landmark will seem just a few minutes away; you'll fly and fly and fly, and it's still just where it was. Don't trust your estimates of distance until you get used to the visibility. Instead, use the chart to relate the airplane's present location to the landmark's location, and determine the actual distance based upon that relationship.

In contrast to flatlands, where weather systems tend to be large and slow moving in the summer but small and fast in the winter, the exact opposite is true in the Rockies. Summer brings relatively small and fast moving systems through the Continental Divide's mountain range; winter and early spring might produce large areas of clouds and precipitation that often hang around for days—great for skiers, but tough for a pilot who is trying to keep a schedule.

Because airports and weather reporting stations are sparse in the mountains, the small summer disturbances might go unreported for a significant period of time. *Don't try to fly underneath or over clouds, and NEVER fly inside mountain clouds.* Circumnavigate anything that even begins to look doubtful. The tops of cloud layers in the

Rockies will exceed the climb capabilities of any normally aspirated airplane. Getting caught on top has dealt a fatal blow to many a VFR pilot in the mountains. You can't go up, and going down means a probable encounter with the unforgiving cloud known as "cumulogranite."

Trying to go underneath a low cloud deck is just as unrealistic. The ceilings will often cling to the ridges, leaving no room to fly between the clouds and the rocks. *Don't try to fly through a valley unless you can clearly see the other side and what lies beyond.* What looks like a valley might be a box canyon with no way out if there are clouds above.

Mountain thunderstorms build with great regularity during the afternoon in the summer and fall. They rarely combine into squall lines. Instead, they rise to stupendous heights in solitary splendor. They are to be avoided at all cost just like any other thunderstorm because they contain all of the turbulence, rain, hail, and vicious winds as their siblings elsewhere. But because they are usually clearly defined, they can most often be circumnavigated, assuming that the terrain and fuel supply permit.

# WIND

Mountainous terrain greatly affects the velocity and direction of winds. When air moves across mountain ranges at any appreciable speed, it usually remains smooth on the upwind side; the velocity will increase at a steady rate as the air is pushed upward and might double by the time that it reaches the ridge.

When flying within the flow of air toward a mountain ridge, a pilot is likely to experience strong and continued updrafts that might cause the airplane to climb several thousand feet without any added power. That's great for an experienced sailplane pilot who is doing *ridge soaring*, but it's trouble for everyone else.

On the other side of the ridge, the wind will tumble down like a giant waterfall, producing considerable turbulence and downdrafts. When the airplane is approaching a ridge line, be ready to increase power and stay high to give you an extra margin of safety against the downdrafts. Many mountain fliers prefer to cross ridges at a 45° angle to minimize the adverse effects; other pilots see that maneuver as only prolonging the misery.

If the wind is at all strong, use the 45° technique because it gives you more time to diagnose just how bad the ridge effect will be, and it will only take a 90° turn to get you going away from the ridge again. If you fly straight at the ridge in strong winds, the updraft will be encountered quickly, and it'll necessitate a 180° turn to get out of there.

Mountain passes and narrow valleys might produce a venturi effect, increasing the speed of any significant wind to dangerous levels. The only alternatives are to either climb above the ridges where the effect is minimized or don't fly through either one.

Flying into a venturi in a slow airplane might result in little, if any, headway, and an encounter with significant turbulence. If you are going downwind with a venturi flow, your groundspeed will increase dramatically, and you could find yourself in a position where you don't have the time to turn around before you encounter a canyon or valley wall.

When approaching a ridge from the downwind side, it's a good idea to do so at a 45° angle to the ridge. Even though this technique will prolong the exposure to the downdrafts and turbulence, escape is much easier if the need arises. By using the 45° method, you can experience a more gradual growth in the intensity of the adverse consequences, and if a retreat is necessary, it takes only a 90° turn, rather than a 180, to get out of harm's way.

Don't play around with the downdrafts on the lee side of a ridge. If you find yourself applying full remaining power without a nice, positive rate of climb resulting, get out of there. Things will only get worse as you get closer to the lee side of the ridge, so don't wait until it's too late to solve the problem with the only available means, which is a turn away from the ridge.

If you do get caught in a prolonged downdraft, don't pull up the nose. Know your aircraft's best-rate-of-climb and best-angle-of-climb airspeeds. Raising the nose beyond what is needed to achieve the desired airspeed will result in worse climb performance, not better. If the turbulence gets bad, go to the airplane's turbulence penetration speed, if published in your manual. If not, be sure you don't exceed the maneuvering speed.

When flying inside a valley where the wind is blowing at right angles to it, fly on the downwind side of the valley with the upwind side of the mountains along the side of the valley nearer to you. The continuous updraft here will increase aircraft performance and simplify things if you have to turn around. By turning into the wind, your groundspeed will be slower, giving you extra time to complete the turn within the confines of the valley.

During the hot summer months, plan your flying during the early morning hours. First of all, no thunderstorms will have likely formed yet, and those from the previous evening will have dissipated. The thermal activity and related turbulence will be at its least intensity. Mountain flying can get rough later in the day, straining not only the comfort of the pilot and passengers, but also the performance and even the structure of the aircraft.

The next better bet is to fly during the late afternoon or in the early evening when the air will be cooler than the middle of the day and there is still plenty of daylight remaining. This isn't as good as the early morning because it might take several hours for the convective turbulence of the hot afternoon to abate, and intense thunderstorms can hang around for quite a while, even after sunset.

## NIGHT FLYING

Night flying in the mountains is only for the experienced mountain flier. Nothing is blacker than a moonless night in the clear air over the endless stretches of uninhabited rocks. The mountains are hard to see, and unless you can fly high enough to safely clear anything in the vicinity, you might not see an obstacle until too late to take evasive action.

The thought of an engine failure in the mountains is terrifying even during the day. Reasonable places to land are essentially nonexistent, but at least in the daylight, you can keep the airplane under control and avoid the worst. At night, it's a crap shoot. I just plainly would not fly in truly rugged mountain areas at night in a single-engine airplane.

# CHARTS

It's imperative to have current sectional charts along for any flight into the mountains. If you haven't practiced your skill at reading the terrain information on them recently, do so before venturing into the mountains. In many situations, it's necessary to quickly determine what's below and around you, without undue time to study the chart in flight.

All sectionals have contour data displayed by chart colorations coupled with the printed elevation of prominent points, obstacles, and airports. Because the terrain changes are marked and take place in very short distances—the vertical face of a cliff is a prime example—learn how to read the chart before you get into a situation where you need that information instantly and do not have the time to mull it over.

# PLANNING AND PREPARATION EQUAL ENJOYMENT

Flying the mountains can be breathtakingly beautiful, but it must not be taken lightly. Sloppy flying skills, slipshod planning, lack of in-depth knowledge of the airplane and its performance limitations, or inadequate maintenance of the airplane, its systems, or electronics can turn a mountain flight into a final flight. Pilots who live and fly regularly in the Rockies and the Appalachians have the utmost respect for the terrain and weather and wind, plus they are intimately familiar with their aircraft. With adequate planning and preparation, you can join them in many happy flights through some of the most scenic territory on Earth.

## Mountain flying tips

- Plan an en route stop at one of the foothill airports prior to entering mountainous terrain.
- Consult a local accident prevention counselor for advice on routing and other factors.
- Check the weather over your entire route. Do not attempt the flight if winds aloft near the mountain tops exceed 40 percent of the aircraft's stalling speed. If weather is marginal, delay the trip.
- Plan trips during the early morning or late afternoon hours.
- Use current charts, preferably sectionals or state air navigation charts. (Use current charts that feature the best possible terrain relief.) Radio navigation might prove difficult due to high terrain.
- Route your trips over valleys wherever possible.

- Learn as much as possible about the airport at your intended destination.
- Carry enough fuel to make your trip with ample reserve.
- Know your aircraft's performance and limitations.
- Make proper corrections for the effects of barometric pressure and outside air temperature when taking off and climbing.
- Check weight and balance of loaded aircraft before takeoff.
- Your normal horizon is near the base of the mountains.
- Beware of rapidly rising terrain and dead ends in valleys and canyons.
- Downdrafts and turbulence occur on the lee side of mountains and ridges.
- Approach a ridge at an angle so you can turn away if you encounter a downdraft.
- Maintain flying speed in downdrafts.
- Carry survival equipment. Even summer nights are cold in the higher altitudes.
- Be prepared for downdrafts and turbulence on final approach.
- Use power-on approaches.
- File, open, and close a flight plan.

(Tips were based upon an FAA list of mountain flying suggestions.)

# 13
# Overwater flying

MOST PILOTS WHO FLY ONLY IN THE CONTINENTAL UNITED STATES MIGHT never need to fly over water for any extended period of time. But if you've ever thought of touring the Bahamas or elsewhere in the Caribbean, you will find yourself over the ocean for quite a while. Sometimes even over the 48 states the most direct route is along a coastline or over large bodies of inland water such as the Great Lakes.

Whatever the reasons for flying over water, there are additional things to know and precautions to take. If you're going to do any serious ocean flying, you'll need to read more than what is contained in this chapter. Get some good books that are solely devoted to the subject, such as TAB/McGraw-Hill's *Ocean Flying—2nd edition*, which is another book in the Practical Flying Series.

First of all, if you are one of those pilots who never files a VFR flight plan for normal cross-country flights, think again about overwater trips. If your route takes you into the coastal or domestic ADIZ/DEWIZ, filing a defense VFR (DVFR) flight plan is an absolute legal requirement. If you don't, you'll be intercepted by fighter aircraft, the Drug Enforcement Administration, or the U.S. Customs Service. After they lead you to a landing, you'll get to answer a lot of questions in a setting where you'd rather not be, and you might just pay some penalties that will ruin not only your whole day, but your entire trip.

While we aren't covering customs procedures in this book, be sure that you're up-to-date on them before leaving or returning to U.S. airspace. Back in the early 1970s, I owned a Piper Aztec that I flew to the Bahamas. When I entered the United States, a customs inspector at West Palm Beach wanted to see inside the nose baggage compartment of the airplane.

Because radios in those days were not solid-state, many of the bulky amplifiers and other electronic equipment were installed in the nose, making the baggage compartment practically unusable; therefore, I didn't carry a key for that compartment on my normal key ring. Don't ask why. If I needed to get into that area to repair a radio, I would have had a real problem. I was about to have a worse problem when I told the customs fellow that I didn't have the key.

He gave me the alternative of finding it or drilling the lock so he could do his looking around. If the lock had been drilled, I couldn't have departed without fixing it to keep the baggage door closed in flight. Try to find a mechanic and the necessary parts to do that on a Sunday afternoon. As is normally the case, my wife saved the day when a thorough search of her purse found the wayward key. (She thinks ahead and brought all of the keys with her. I haven't complained about packing the kitchen sink on trips since then.) The inspector satisfied himself that I was not a drug runner, and the Aztec was airborne after some consternating moments.

Even in areas where flight plans aren't legally mandated, always file one for any overwater flight. If you do have to ditch, time is your enemy in any but the warmest water. The sooner the search starts, the more likely you are to be found alive. The search and rescue system will be activated if you don't close your flight plan within 30 minutes of your ETA.

A commonsense rule for short overwater flights, or for those along coastal areas, is to fly high enough to be within gliding distance of the shore. This isn't always possible. The weather might prevent operating at high altitudes, or such a height might be beyond the capabilities or your airplane, or you might not have oxygen on board.

Another concern is whether the shoreline is suitable for a forced landing. If it's rocky or littered with cliffs, you might have a better chance in the water than you would on land. The only way to mitigate against the dangers of ditching is to carry emergency survival equipment, and learn a little bit about ditching techniques.

The other obvious option is to avoid the situation whenever possible and remain inland even if the distance of the inland route is farther than a direct route over the water. Flying is an activity that is replete with risk assessment and acceptance. Use your own judgment, but use it wisely.

## OVERWATER SURVIVAL EQUIPMENT

For some reason known only to the bureaucrats, the FARs don't require any overwater survival equipment for flights in single-engine airplanes operating under Part 91. FAR 91.509 does set forth requirements for large and turbine-powered aircraft, and if the flight is planned for more than 30 minutes or 100 nautical miles from shore, the list of required equipment is extensive.

This section of the regs is a bit ironic; when was the last time that a bizjet had to ditch, compared with light singles? Nevertheless, FAR 91.509 is a good guide if you make an overwater flight in any airplane.

The regulation requires that flights over the water more than 50 nautical miles from the nearest shoreline have a life preserver or other flotation device carried on the airplane for each occupant. That should be the minimum that you carry, even though the regulation doesn't apply to lightplanes. Even with a life preserver, 50 miles to the nearest shore is an impossible distance to swim. For that matter, so is 3, or even 1 mile.

When was the last time you swam a mile? Unless you're a triathlete, you probably haven't done it since you were a teenager or in the military, and then it was probably in a pool. Swimming even a mile in the open water of the ocean or a large lake is quite another feat.

For anyone contemplating flight over any significant body of water, a life preserver (vest) and a life raft are essentials if you are going to have a snowball's chance of surviving a ditching. What constitutes a flight over a significant body of water is any flight over water beyond the combined distance that the airplane can glide to land and that you or your passengers can swim.

Even if you're an excellent swimmer and enjoy the physical condition of an Olympian, hypothermia can and will kill you. If the water temperature is anywhere below your body temperature (98.6°F average), you'll start losing body heat the moment you hit the water. The thermal conductivity of water is 240 times greater than still air. This means that water or wet clothing will extract body heat up to 240 times as fast as standing on dry land. The colder the water, the worse it gets as you lose body heat very rapidly. For that reason, if you're going to have to spend more than a few minutes in the water before you can get to dry land, you have to get into a raft.

Figure 13-1 describes survival time in the water. Even in water warm enough to be comfortable for a recreational swim, you'll be dead in 12 to 15 hours. If you're swimming around in a life vest in the middle of the Caribbean, it can easily take 12 to 15 hours, and maybe much longer, to be rescued.

Some people talk about taking survival suits along for flights over cold water. That's great if you can stand to wear it throughout the flight, but most people would not be able to do so. There is little chance that you'll be able to don it after ditching. Don't try to fool yourself with unreasonable solutions for a deadly risk.

Unless you are a true high-stakes gambler, it's foolish to fly a lightplane over water in winter climates. I wouldn't even think of flying across one of the Great Lakes in winter, except maybe Lake Erie, which is so shallow that it usually freezes solid all the way across, and cars are driven out to the ice-fishing huts.

Another important reason to use a life raft is that a raft is much easier to spot from a search airplane than is a person afloat in a vest. The single person in a life preserver is nearly impossible to find amongst any waves unless the search aircraft flies right overhead. If there are passengers in the airplane with you, it'll be next to impossible to stay together after a ditching unless you all get into a raft.

In a short time, maybe as little as a few minutes in high seas, you'll lose sight of

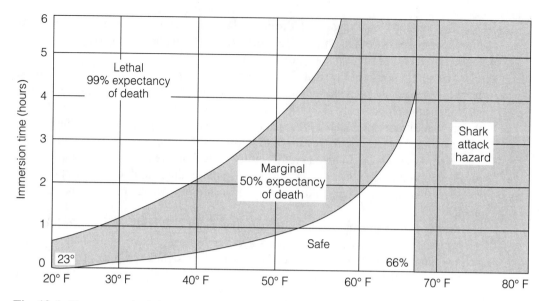

**Fig. 13-1.** *Human survival time in open water.*

each other if you're floating in life vests. Staying together in a raft makes it much more likely that everyone will get rescued. If the temperatures are cold, being together in the raft also helps you keep warm.

Remember that wet clothing will hasten the onset of hypothermia. Keeping warm and dry are two of the biggest keys to surviving a ditching when you have emerged from the airplane relatively unscathed.

The next item to have is some sort of signaling device. A person afloat in a vest or in a raft will see an oncoming aircraft or boat long before anyone in the plane or boat sees someone waiting to be rescued. The chances of being seen are greatly improved if they can be signaled.

The best device available today is the hand-held communication radio, which can cost $500 and up (1994 dollars). But don't depend solely upon a radio because it can get wet and fail, and the batteries might not be fresh, despite your best intentions of always having good batteries. If you can wrap it in a waterproof wrapping, all the better. That covering can always be removed if you want to use the hand-held for normal communications while airborne.

Remember that the international emergency frequency in the VHF band is 121.5 MHz. Search aircraft should be monitoring it. Less expensive devices include signal mirrors, flares, and sea dye. Some life rafts come equipped with survival kits that include all three; buy them, if necessary.

Shiny, metal signal mirrors are great. They don't break, don't have batteries, don't have to be kept dry before needed, and don't require special training prior to use. They reflect light toward the search aircraft or boat in almost any daylight, unless the sun is totally obscured.

Remember that in a pinch any piece of shiny substance, even glass, can probably be used as a signal mirror. Take a look at Fig. 13-2 for the recommended way to use a mirror to attract attention.

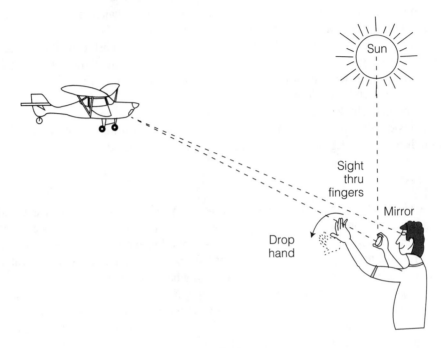

**Fig. 13-2.** *The best way to use a signal mirror.* Tacoma Mountain Rescue Unit

Flares come in day and night styles, but each can be used anytime. A day flare puts out more smoke than bright light, and a night flare emits an intense light. If you've used up all of one type, using the other anytime is better than nothing at all. Even a day flare produces some light that can be seen at night.

You can also buy either hand-held or projectile flares. The projectile types are generally fired from a device that looks very much like a pistol; the airborne flare can be seen for miles, especially at night. A hand-held flare is better to pinpoint your exact location. That's why it's best to have both.

Signal a distant aircraft or boat with the pistol flare, then use the hand-held flare if a rescuer appears to be headed toward you. Smoke from a hand-held flare can be a tremendous help to a helicopter crew in the daytime to aid them in determining wind velocity and direction.

Sea dye makes such a big "spot" in the water that it's more effective than most people would think. Depending on the state of the sea, it will probably last longer than a flare. The dye has even been thought to be a shark repellent in some instances. Don't use sea dye until you see either a large ship or an aircraft approaching your position.

Don't waste the dye by using it when the ship or aircraft is so far away that the crew doesn't have the slant-range visibility to plainly see the water's surface. As with any signalling device, the approaching rescue crew has to be able to see the discolored water before the signal does you any good. Hasty use of these devices before they can be seen by the approaching ship or aircraft just reduces your chance of being rescued because you've thrown away one more chance of being seen.

Sharks are another reason to have a life raft. While a big shark can attack a raft, you have a much greater chance of becoming a target if you're treading water in a life vest. Sharks are particularly attracted to the smell of blood; if you're injured in the ditching, even slightly, you'd better be in a raft.

Like people, sharks prefer warm water, but in cold water, your time is severely limited outside of a raft. That ought to tell us to use a raft regardless of the water temperature if we want a fighting chance at surviving a ditching at sea.

# DITCHING

Ditching any airplane, under any conditions, is tough. You certainly can't practice it, and your adrenalin will probably be pumping as never before. If you have to ditch, accept the fact that you are simultaneously going to be a student pilot learning something new and an experimental test pilot trying something for the first time. Manufacturers don't ditch airplanes in testing, so nobody is around to tell you what it is going to be like.

Prepare for a crash, and do your best to keep your cool. Many pilots and passengers have survived ditchings, especially in warm water. If you do your best to keep the airplane under control, touch down properly, and get out of the airplane into a life raft, your chances of survival are greatly enhanced.

While still at altitude, start making Mayday calls on 121.5 MHz. Another good reason to have either a loran or GPS is that you can determine your present position in an instant and relay it, even in the blind, during those calls. Don't forget to clearly state that you're ditching.

Attain the slowest speed and rate of descent that are consistent with safe control of the airplane. Water is like concrete when impacted at high speed, and impact forces increase exponentially with the increase in touchdown speed. Do your best to contact the water in a stall. Just a few knots difference in the speed at which you contact the water might save your life.

I don't fly over water without always wearing the life vest. Uninflated, the vest is not that uncomfortable. If a naval aviator can wear one all of the time, I can for the few times when I fly over water.

There isn't time to put it on after the engine quits, unless you have no other choice. Never inflate it inside the airplane, or you won't ever get out. Watch for the disorientation that will most probably set in right after the airplane comes to a stop, and make sure that you are all of the way out of the aircraft before inflating your life vest.

During the descent to the surface of the water, you have to try to get some idea of the wind direction and state of the sea. Neither is easy, unless you're an experienced naval aviator or sailor. The AIM goes into great detail about oceanographic terminology, most of which is meaningless to us landlubbers.

Waves reveal the condition of the surface of the sea caused by local winds. A wave that is left over from a distant disturbance or storm is called a *swell*. More than one swell condition can exist in one patch of water. Wind-created waves can be going in different directions than the primary, secondary, or even tertiary swells. Add all of this to the fact that there are riverlike currents flowing in the ocean, and the entire situation becomes extremely complex.

The main point to remember is to avoid hitting the face of a swell, which is the side of a swell toward the observer. In other words, don't land into a wall of water. That's much easier said than done. When the wind is from one direction and the swells from another, you'll be making a crosswind landing onto a surface that is simultaneously moving horizontally and vertically. As the AIM recommends, try to land on the backside or crest of a swell. Regardless of the initial contact, you'll be facing another swell as the airplane decelerates, and the nose will plow in, usually somersaulting the airplane.

If you're flying a retractable-gear airplane, be sure to ditch with the wheels up to increase your chances of staying upright and to lessen the forces of impact. In a fixed-gear airplane, be even more prepared to go inverted as the airplane digs into the sea. Unless you really need them, avoid the use of wing flaps. It is very common for the impact to tear away one flap, causing the airplane to skew as it decelerates. That adds further to the chances of injury, upset, and disorientation.

Try to open a door or window prior to impact. If you don't have a chance to do so, as soon as the airplane starts going under, the pressure of the water will make it virtually impossible to open a door if the cabin is still closed. Structural failure might also jam doors and windows shut. Open or break a window to allow enough water into the cabin to equalize the pressure enough that you will be able to open the door. Grab your life raft and survival kit, and exit the aircraft, ensuring that any passengers are also able to exit safely. Once safely outside of the airplane, inflate your life vest and continue to assist any passengers.

Most life rafts will float even before they are inflated. Check yours in a swimming pool to be sure. If it doesn't float without inflation, take it back. If it starts to sink before you can inflate it, you not only can't use it, but you could be pulled downward by it as you try to swim toward the surface after exiting a sinking airplane.

If it does start to sink, pull the inflation lanyard after you're safely out of the airplane (never before), and tightly hold on to the cord. An empty inflated raft is extremely buoyant and lightweight; strong winds will blow an unattended and empty raft across the surface of the water much faster than you can ever swim after it. Hold that cord very tightly. Climb into the raft, get your signaling devices ready, and check your hand-held radio before it gets totally soaked. Be thankful that everything is going as well as can be expected and wait.

Doing this maneuver right the first time without ever having practiced it is going to take a good measure of luck or divine intervention. If you plan much overwater flying, go to one of the relatively inexpensive schools that provide underwater egress training. That's about as close to a real ditching as anyone with his or her senses will want to ever experience. Remember that all Naval aviators get a ride in the "dunker" early in their training. Your doing likewise might make the difference in favor of surviving a ditching.

Figure 13-3 displays a drawing from the *Airman's Information Manual*. It shows many of the variables that come into play when an airplane is ditched. Take time to study it now, and consider putting a copy with your charts so you have it ready for reference if the need ever arises.

## FUEL

There is nothing more important than having enough fuel for overwater operations. Most overwater flights in lightplanes, especially over island groups, don't stretch the fuel capacities of the aircraft if the fuel tanks are full. There is no reason to ever "push" adequate fuel and proper fuel reserves when flying over water.

If the weather "goes South," you'd better have enough gas to go back to your departure point, regardless of how close you were to the destination when you had to divert. There should never be a point of no return during an overwater flight in a lightplane, unless you're an experienced ferry pilot. Always plan the flight to be able to go back to your point of departure from any point in the flight, even the destination airport.

A destination airport might be weathered in, have just experienced an accident that closed it, or a myriad of other catastrophes might happen to deny you its use. Casual recreational flying never justifies the risk of not being able to go back to where you began.

Many people have flown their lightplanes across the vast oceans with wonderful outcomes. Some have not. The degree of risk that you and your passengers are willing to accept is a purely personal decision. But if you even think about such an adventure, prepare for it with intense study and planning.

Talk to ferry pilots who have done it many times in the past. I have known pilots who have flown light singles across the Atlantic and Pacific, but when given many opportunities to join their ranks, I've repeatedly declined. In the final analysis, it's up to you.

Shorter trips over water can be outstanding cross-country flights. The Caribbean is probably the most popular destination for recreational overwater fliers. Personal trips have proved that a dose of Bahamas sunshine really cures cabin fever during the winter months. Be careful, carry the right equipment along, and plan your flight conservatively. You'll remember the time forever.

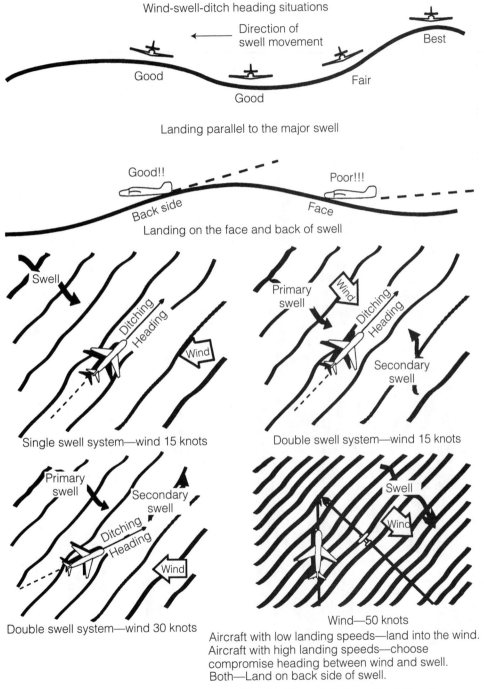

Wind-swell-ditch heading situations

Direction of swell movement

Best

Good

Fair

Good

Landing parallel to the major swell

Good!!

Poor!!!

Back side

Face

Landing on the face and back of swell

Swell

Ditching Heading

Wind

Single swell system—wind 15 knots

Primary swell

Wind

Ditching Heading

Secondary swell

Double swell system—wind 15 knots

Primary swell

Secondary swell

Ditching Heading

Wind

Double swell system—wind 30 knots

Swell

Wind

Wind—50 knots
Aircraft with low landing speeds—land into the wind.
Aircraft with high landing speeds—choose
compromise heading between wind and swell.
Both—Land on back side of swell.

**Fig. 13-3.** *Ditching techniques as prescribed by the* Airman's Information Manual.

# 14
# Ecstasy

THIS BOOK HAS TRIED TO COVER SOME OF THE EQUIPMENT, HAZARDS, AND unexpected challenges involved in cross-country flying. Coupled with those elements are the pure joy, fun, and euphoria that comes with any well-planned and executed flight, especially when the trip takes you to new and different places.

A personal airplane is a wonderful tool for business and a magic carpet for pleasure flying. There are really only two countries in the world, the United States and Canada, that have the affordable freedom to fly lightplanes. Europeans can fly, but only individuals with the highest incomes can afford it. Landing fees approaching $50 are common at most good-size European airports, and avgas pushes $4 to $5 per U.S. gallon. An airplane like a Cessna 172 or Piper Cherokee rents for around $150 an hour in most of the western European countries. The air traffic control system charges for its services, and the costs aren't cheap. The amount of restricted and prohibited airspace in Europe would astound an American or Canadian pilot. For these reasons, a good deal of the private, recreational flying in Europe is done in sailplanes, utilizing winch tows to save the money that a towplane would cost.

Aviation is not quite so restricted or expensive in Mexico and the other Latin American countries, but due to the depressed economic conditions there, very few people indeed can participate in general aviation because of the tiny percentage of the population that has incomes anywhere close to the average American. We in North America are truly blessed in many facets of our lives, especially if we like to fly.

I learned to fly in the summer and fall of 1965, right after I graduated from high school. I was one of those kids who lived to fly airplanes from the time that I was about 12 years old. In my school years, a kid got around $1.50 to mow a yard or shovel the snow from sidewalks and driveways in my old neighborhood. In those times, primary dual instruction in an Aeronca Champ or J-3 Cub ran about $8 to $10 dollars an hour, pretty evenly split between the aircraft rental and the instructor's fee.

Each lawn mowed or few hours spent with a snow shovel equalled a few precious minutes in the sky. By the time high school ended, I had enough money saved to at least solo, which I did after about 6½ hours of dual in a Champ. The airport from which I learned had a single runway that was 2,600 feet long. The field had a gravel surface, fairly low power lines at one end, and three broadcast radio towers that stuck up above the pattern altitude; I had to fly *around* the towers on downwind. I learned how to slip (Champs don't have flaps), make crosswind landings, and avoid obstacles.

Then after renting the Champ for a few more hours, I came to the brilliant conclusion that the way to afford this endeavor was to buy an airplane in which to build the time needed for a private certificate, and maybe even a commercial.

My widowed mother wasn't a happy camper when I took $600 from the small college fund that my parents had established for something meaningful, like an education, to buy a one-half interest in an old "canvas" airplane. The door latch was so weak that the door popped ajar during most takeoffs, including my mother's first ride with me from the rough gravel runway where I was learning to fly. Previous chapters of the book related a story or two about flights in that Taylorcraft; I won't bore you with all my other stories, but I'll never forget them or that airplane.

After a year of flying at the airport where I first learned, I had enough time to qualify for a commercial and CFI. Thanks to the opening of the first Sears store in Columbus, I was able to get a part-time job that paid enough for tuition at Ohio State University, a driveable VW Beetle, and flying. You didn't have to have an instrument rating or complex airplane time in the 1960s to get a commercial or flight instructor certificate; therefore, it was far less costly to progress in the ratings game than it is for today's aspiring pilots.

Then in late 1966 I sold the T-craft and went across town to another airport that was operated by a true icon of general aviation, Dick Moss. He took me from a young kid who had fewer than 300 hours to a real pilot with those coveted commercial and CFI certificates. I spent almost the next two years sitting in the back of another Champ, giving dual and teaching people to fly.

The two years flying for Moss also resulted in a multiengine rating and the chance to fly some airplanes that many pilots can only read about: AT-6, Cessna UC-78 Bamboo Bomber, Howard DGA, Stinson 108-2 and 108-3, Aeronca Sedan, and Cessna 140 and 180. An early model Cessna 150 was available for those students who had the big bucks and wanted to fly the newest and best trainer available.

While working with Moss, I got to fly banner tows and do quite a bit of aerial photography survey work. The camera installed in the rear passenger area of the Cessna 180 that Dick owned was worth more than the airplane. Because it had a glass viewing port in the belly, you had to do wheel landings all of the time to avoid kicking gravel from the runway into it and breaking the camera lens. I still hadn't gotten away from gravel runways yet.

Those were great years, and I wouldn't trade the time I spent at that airport during college—which was every waking hour not spent in class, and a good number of hours that should have been in a classroom—for all of the fraternity parties and campus events in which my buddies took part. I passed up those extracurricular activities to spend time instructing and flying.

Dick decided to close his operation at Columbus Airpark in 1968 and moved to Florida. I was subsequently hired at the Ohio State University Department of Aviation as a staff flight instructor. If you've ever experienced true culture shock, you might know what is was like to go from the informal setting of single-runway gravel fields where we taught our students in Champs to the structured training environment of a major university's flight department. I'm not sure that the latter produced any better pilot than the former, but that subject and the debate over it is for another time.

I was able to cajole one of the other OSU instructors into giving me the dual for an instrument rating, which I finally got when I had about 2,500 hours in my logbook.

I met a pretty little lady at the OSU Airport who was another instructor's flight student. Thankfully, she wasn't my student, or she probably would have never become Mrs. Eichenberger shortly after we graduated. I met her father for the first time when when he returned to Port Columbus International Airport from a business trip. His daughter and this kid whom he had never seen before showed up with a Cessna 172 to fly him all of the 30 miles to the town where he lived.

He was gracious enough to accept the ride in the middle of the night even though it meant leaving his car at the large airport and going back the next day to retrieve it. My future mother-in-law came to their little town's small airport to get him, probably wondering what her first-born daughter was dragging with her. Five years after he gave me his little girl at the front of the church, I taught him to fly.

I met a great number of people at OSU who have remained my best friends. The best man at our wedding and his brother, who was an usher, were both flight students. I still talk to them several times each year and see them during football season for a few games when we all still sit in the big horseshoe stadium where the Buckeyes try to re-gain the glory years of Ohio State football. Both are very successful today, one as a businessman, the other as a fellow lawyer. Two other ushers were guys whom I had taught to fly, both of whom were successful in business when I met them and continue to be so today.

While on the staff at OSU, I got to fly some more venerable airplanes, such as the DC-3. Today's youngsters in aviation have the chance to fly jets, but we dreamed of Howard 500s and Aero Commanders as the pinnacle of corporate aircraft. I remember the day that the first Learjet came to OSU Airport to be based there. I also worked part-time during college selling and ferrying airplanes all over the country.

My first flight through the Rockies was in a Cessna 180 that I had brokered and was delivering to its new owner in Oregon. He had no tailwheel time, and no matter how hard I tried, I couldn't convince him to come to Ohio to take delivery, even with the promise of enough dual in the airplane until he felt comfortable going home in it. You read about that delivery, when the crosswind-gear saved the airplane on that windy day in Colorado.

After the T-craft flight to Florida, the next time that I flew down to the Sunshine State was in a PA-11 that I was ferrying for a businessman who had moved to Florida from Ohio and didn't want to fly that far in a plane without radios. The last really long trip that he had made was flying a B-17 back from his last mission in Europe at the end of World War II. My personal trip to Florida had taught me well, and I didn't get lost on the ferry trip through the mountains in the PA-11.

After OSU, it was on to law school and a "real job" as a corporate pilot for a real estate development company. Sometimes I wasn't seen in law classes; the demands of flying came first because I had to put food on the table first and worry about school second. I made my first overwater flights to the Bahamas because the owner of the company loved to go over there every time business took us to Florida to oversee a building project that was in progress.

After law school, I was fortunate to combine my avocation and professional vocation by devoting my law practice primarily to aviation and real estate. That blessed mix of fun and work hasn't ended, and I sincerely hope that it doesn't. Most lawyers spend their careers either in courtrooms trying all sorts of cases, or in an office practice drafting documents and closing deals. I get to do both to some extent, although I spend most of my time in aviation litigation, where I get to combine what I love best about both aspects of my working life.

During the first few years that my wife and I were married, Dick Moss came back from Florida to Columbus and established another small FBO at a rural airport in southern Ohio. My bride got to meet him and flew with the master who could fly the crates that airplanes came in.

In all the years that I've been a CFI and check pilot, I've never seen anyone who could fly like Dick Moss. You never noticed that he was moving the controls; the airplane just did what he willed it to do. I can't ever remember a bad landing with him in the cockpit, whether he was flying the AT-6 or a Champ. He died in the late 1980s. Every pilot has some mentor in his or her career who taught him or her to be an aviator, not just an airplane driver. For me, it was Moss.

I cannot begin to recount all of the cross-country flights that my wife and I made before and since we had a daughter. We've been all over the United States and the Bahamas, flying through murky and brilliant skies at night and in the day. Each flight was different and memorable. When you enjoy flying with your family, everybody has those great times that memories are made of.

Part of the family has taken up glider flying. A person can get a student license and solo a glider at age 14. All three of us went out to the flight standards district office when my daughter got her student certificate on her 14th birthday, accompanied by dad's camera and all. I arranged for one of the women inspectors to conduct the interview, take her application, and issue her the student license. She and I learned to fly gliders together.

Mom wasn't too keen on sailplanes, preferring instead to have a Continental or Lycoming keeping her aloft, rather than going out to hunt thermals. Husband and daughter are still working on her to get the add-on glider rating. Maybe persistence will pay off; if it doesn't, we'll think of something else to get the job done. Flying a glider is one whale of a lot of fun, and it will teach any pilot the finer points of energy management and planning ahead, which hones skills and improves safety.

When we talk about the pleasant times, as every family does, most of those chats revolve around something to do with flying. We remember several cross-countries we've taken together with fondness. The year before our daughter was born, my wife and I took our Cessna 150 from Ohio to Texas, via Iowa. We droned through the Midwestern sky for 33 hours, and even though I've flown thousands of hours since, the memories of that trip will always be with me.

The day after our daughter was born, while she and her mother were still in the hospital, I flew our Cessna 170 about 60 miles to announce the event to some good friends who have their own private strip. Our new arrival took her first airplane ride when she was two weeks old. She sat on her mother's lap at age 3 with hands on the control wheel looking around with a grin on her face as the Cessna 170 would dance up, down, and around from her control inputs. She still loves a negative-G parabola. That 170 took our little bundle of joy to see my mother for the first time because Grandma had moved away from the Columbus area by then.

From the time our little girl was 4 years old until she was 11, we had a Piper Comanche. The problem then was that she wasn't tall enough to reach the rudder pedals and look out of the windshield at the same time; she could do one or the other, and you don't enjoy it much when you're scrunched down under the panel. So when dad taught her to take off, I worked the pedals, and she flew the rest of the departure. Landings were even more challenging, but we never broke anything.

After a two-year hiatus from airplane flying, when I mostly flew helicopters, I got into sailplanes. Imagine a father's feelings the first time that he flies a tow plane with his daughter as pilot in command at the controls of the glider that is behind him.

Each step in our aviation lives has been a new challenge, and the most fun we've had. I truly pity all of those people on the ground whose homes and farms we've flown over, who have never experienced what we have. It's here for the taking, and I can't imagine why anyone wouldn't love to take advantage of it.

Go out and fly some cross-countries. Take your friends and families with you. If they're initially cautious or fearful, that's natural. Let them get used to lightplanes gradually. Always explain what you're doing. Show them that the reason you're checking the mags during the run-up is that airplane engines have two independent ignition systems, and you're just proving to yourself that the airplane will run just fine on either one.

Fly aviation newcomers on smooth days, early in the morning or in the evening, when the air is smooth and the experience will be exhilarating. Night flights are fascinating, so introduce them to the starlit environment of the sparkling lights on the ground and in the sky. If you live in an area of the country that has a winter season, let them see their world under a blanket of snow.

Final analysis: Enjoy what personal flying is really all about—ecstasy.

# Index

Illustration page numbers are in **boldface**.

# About the author

The author of *Cross-Country Flying*, Jerry A. Eichenberger, has written two other books in the TAB Practical Flying Series. *General Aviation Law* was written in 1990, and he revised and updated the 5th Edition of the perennial bestseller, *Your Pilot's License*, in 1993. He has written approximately 130 magazine articles about general aviation law, piloting techniques, and aircraft pilot reports.

Eichenberger is an aviation attorney and is a partner in one of the oldest Columbus, Ohio, law firms: Crabbe, Brown, Jones, Potts, and Schmidt. He learned to fly in the summer of 1965, right after high school graduation. He worked in the general aviation industry as a charter and corporate pilot during college and law school days; Eichenberger held a flight instructor certificate for 22 years.

He has owned a classic Taylorcraft, Beechcraft Bonanza, Piper Aztec, Cessna 150, Cessna 170, and a Piper Comanche. Eichenberger has flown virtually every piston-engine airplane made in the United States since World War II. He has logged approximately 5,000 hours in aircraft from the Piper Cub through transport category airplanes, plus piston- and turbine-powered helicopters, and sailplanes.

Eichenberger is a commercial pilot with ratings for single-engine and multiengine airplanes, helicopters, and gliders, plus airplane instrument privileges.

His cross-country flying experience covers flights throughout the continental United States, Canada, the Caribbean, and the United Kingdom. His wife is also a certificated private pilot and is an aviation insurance broker. His teenage daughter is a student pilot.

# Other Bestsellers of Related Interest

**Bush Flying**
*Steven Levi and Jim O'Meara*
A survival guide for bush pilots, designed to prepare experienced aviators for the challenge of flying under treacherous conditions. This book emphasizes the unique skills needed to fly in remote areas of Alaska, Canada, Scandinavia, and Australia. Steven Levi and Jim O'Meara explore every facet of this unpredictable brand of flight, and by doing so enable readers to forgo mistakes that could prove costly.
**ISBN 0-8306-3462-2, #157163-9    $17.95 Paper**

**Night Flying**
*Richard F. Haines and Courtney Flatau*
This is the first book to approach the subject of safe night flying both as a matter of technique and of personal fitness. No other book combines practical information on the latest aircraft technology with the most current research into the human factors of flight safety.
**ISBN 0-8306-3773-7, #025809-0    $17.95 Paper**

**Weather Patterns and Phenomena:  A Pilot's Guide**
*Thomas P. Turner*
A volume in the TAB Practical Flying Series, this book tells pilots how to assess aviation weather hazards correctly and confidently. Chapters cover weather theory in depth, including specific hazards such as thunderstorms, turbulence, reduced visibility, ice, and distinct regional weather patterns.
**ISBN 0-07-065602-9    $16.95 Paper**

## How to Order

 **Call 1-800-822-8158**
24 hours a day,
7 days a week
in U.S. and Canada

 **Mail this coupon to:**
McGraw-Hill, Inc.
P.O. Box 182067
Columbus, OH 43218-2607

 **Fax your order to:**
614-759-3644

 **EMAIL**
70007.1531@COMPUSERVE.COM
COMPUSERVE: GO MH

### Shipping and Handling Charges

| Order Amount | Within U.S. | Outside U.S. |
|---|---|---|
| Less than $15 | $3.50 | $5.50 |
| $15.00 - $24.99 | $4.00 | $6.00 |
| $25.00 - $49.99 | $5.00 | $7.00 |
| $50.00 - $74.49 | $6.00 | $8.00 |
| $75.00 - and up | $7.00 | $9.00 |

# EASY ORDER FORM—
# SATISFACTION GUARANTEED

Ship to:

Name _____

Address _____

City/State/Zip _____

Daytime Telephone No. _____

### *Thank you for your order!*

| ITEM NO. | QUANTITY | AMT. |
|---|---|---|
|  |  |  |
|  |  |  |

Method of Payment:

☐ Check or money order enclosed (payable to McGraw-Hill)

☐ DISCOVER          ☐ AMERICAN EXPRESS Cards

☐ VISA              ☐ MasterCard

| | |
|---|---|
| Shipping & Handling charge from chart below | |
| Subtotal | |
| Please add applicable state & local sales tax | |
| TOTAL | |

Account No. ☐☐☐☐☐☐☐☐☐☐☐☐☐☐☐☐

Signature _____ Exp. Date _____

Order invalid without signature

**In a hurry? Call 1-800-822-8158 anytime, day or night, or visit your local bookstore.**

Key = BC95ZZA